P9-AZW-861

Serendipity

THE WILEY SCIENCE EDITIONS

The Search for Extraterrestrial Intelligence, by Thomas R. McDonough
Seven Ideas that Shook the Universe, by Bryon D. Anderson, and Nathan Spielberg
The Naturalist's Year, by Scott Camazine
The Urban Naturalist, by Steven D. Garber
Space: The Next Twenty-Five Years, by Thomas R. McDonough
The Body in Time, by Kenneth Jon Rose
Clouds in a Glass of Beer, by Craig Bohren
The Complete Book of Holograms, by Joseph Kasper and Steven Feller
The Scientific Companion, by Cesare Emiliani
Starsailing, by Louise Friedman
Mirror Matter, by Robert Forward, and Joel Davis
Gravity's Lens, by Nathan Cohen
The Beauty of Light, by Ben Bova
Cognizers: Neural Networks and Machines that Think, by Colin Johnson and Chappell Brown
Inventing Reality: Physics as Language, by Bruce Gregory
The Complete Book of Superconductors, by David Wheeler, and Mark Parish
Planets Beyond: Discovering the Outer Solar System, by Mark Littmann
The Oceans: A Book of Questions and Answers, by Don Groves
The Starry Room, by Fred Schaaf
The Weather Companion, by Gary Lockhart
To the Arctic: An Introduction to the Far Northern World, by Steven Young
Ozone Crisis: The 15-year Evolution of a Sudden Global Emergency, by Sharon Roan
The Endangered Kingdom, by Roger DiSilvestro
Serendipity: Accidental Discoveries in Science, by Royston M. Roberts
Senses and Sensibilities, by Jillyn Smith
Chemistry for Every Kid, by Janice Pratt VanCleave
The Atomic Scientists: A Biographical History, by Henry Boorse, Lloyd Motz, and Jefferson Weaver
The Starflight Handbook: A Pioneer's Guide to Interstellar Travel, by Eugene Mallove, and Gregory Matloff

Serendipity

▼ ACCIDENTAL DISCOVERIES IN SCIENCE

Royston M. Roberts

WILEY SCIENCE EDITIONS

WILEY

John Wiley & Sons, Inc.

New York ▼ Chichester ▼ Brisbane ▼ Toronto ▼ Singapore

PUBLISHER: Stephen Kippur
EDITOR: David Sobel
MANAGING EDITOR: Frank Grazioli
EDITING, DESIGN, AND PRODUCTION: Michael Bass & Associates

Library of Congress Cataloging-in-Publication Data

Roberts, Royston M.
Serendipity : accidental discoveries in science / Royston M. Roberts.
p. cm. — (Wiley science editions)
Bibliography: p.
Includes indexes.
ISBN 0-471-50658-3. — ISBN 0-471-60203-5 (pbk.)
1. Serendipity in science. I. Title. II. Series.
Q172.5.S47R63 1989
509—dc19 88-33638
CIP

Printed in the United States of America
89 90 10 9 8 7 6 5 4 3 2 1

▼

To Phyllis

meeting her was

the greatest

Serendipity

of my life

CREDITS

Grateful acknowledgment is made to the following for permission to use the illustrations listed below.

American Institute of Physics Niels Bohr Library: (W. F. Meggers Collection) photo, p. 12; photo, p. 17; (Lande Collection) photo, p. 17; (Lande Collection) photo, p. 141; photo, p. 146; (Burnby Library) photo, p. 145; (Courtesy of Otto Hahn, *A Scientific Autobiography*) photo, p. 148.

National Foundation for the History of Chemistry: photo, p. 26; photo, p. 27; photo, p. 30; photo, p. 60; photo, p. 69; photo, p. 84; photo, p. 93; photo, p. 128; photo, p. 182; photo, p. 184.

***Fundamentals of Chemistry,* by James E. Brady, and John R. Holum, 3d edition, John Wiley and Sons, NY, 1988:** photo, p. 35; photo, p. 63.

The University of Texas Harry Ransom Humanities Research Center: (Courtesy of the Gernsheim Collection) photo, p. 51.

Larry Kolvoord and the Austin American-Statesman: photo, p. 106.

Dr. Prescott Williams, Austin, Texas: (private collection) photo, p. 115.

Donald L. Frey, Institute of Nautical Archaeology, College Station, Texas: (Courtesy of Dr. George F. Bass) photo, p. 118.

E. I. Du Pont de Nemours & Co., Inc.: photo, p. 158; photo, p. 171; photo, p. 172; photo, p. 176; photo, p. 188; photo, p. 189; photo, p. 242.

Squibb Corporation, Princeton, New Jersey: photo, p. 161.

Dr. Daniel W. Fox and the General Electric Company: photo, p. 218.

Velcro USA, Inc.: photo, p. 221; photo, p. 222.

3M Company: photo, p. 224.

Sir Derek Barton: photo, p. 233.

Professor Brown: photo, p. 210.

The author provided the following photos: photo, p. 44; photo, p. 107; photo, p. 142; photo, p. 223; photo, p. 230; photo, p. 230; photo, p. 241; photo, p. 241.

FOREWORD

During my first year of graduate school in 1940 I met with what Professor Roberts calls "pseudoserendipity" for the first time. I was working on a wartime project, the synthesis of vinyl chloride (for making poly(vinyl chloride), now the world's most widely used plastic). At that time, vinyl chloride was made industrially by adding hydrogen chloride to acetylene. I studied an alternative procedure—one in which ethylene dichloride was passed through a hot tube. All the vinyl chloride synthesized nowadays is made by this method.

I originally purified the ethylene dichloride just by distillation, but one day I tried an additional purification step before distillation; I found that the production of vinyl chloride then proceeded much faster and at a lower temperature. But this result was not always reproducible from day to day. It was exasperating! Finally, I discovered the truth. The variability was due, first, to the additional purification step, which removed an inhibitor to the desired reaction and, second, to an unplanned (and unexpected) air leak in the equipment, which allowed the introduction of oxygen, a catalyst for the desired reaction. Thus, the serendipitous removal of an inhibitor and addition of traces of oxygen produced a phenomenon of tremendous importance to the industry. Since 1940 I have frequently made serendipitous discoveries. Certainly such discovery depends on the prepared mind.

This book, by Professor Royston Roberts, is a fascinating account of serendipity in several fields of human endeavor. It is written in a clear, straightforward way, such that it can be understood and enjoyed by those with little or no technical knowledge.

I think that this book is very welcome. Reading it makes one realize how much scientific advance cannot be planned. When you write a proposal for a funding agency, it is based on current knowledge, not on the unknown. Yet the most interesting science is to be found in the unknown world. How do you go from the known to the unknown? The best way, in my opinion, is to back those who have done it before. Discovery, by serendipity or by conception, usually comes to the same people over and over.

This excellent book should be read by people of all ages. However, it will be particularly useful for those at an early stage in their careers. When you are young, you want all new knowledge to fit into the theories of the time ("the dogma of the day"). Happily, this is not the way it always happens in the real world.

Sir Derek H. R. Barton

INTRODUCTION

▼ What do Velcro, penicillin, X rays, Teflon, dynamite, and the Dead Sea Scrolls have in common? Serendipity! These diverse things were discovered by accident, as were hundreds of other things that make everyday living more convenient, pleasant, healthy, or interesting. All have come to us as a result of serendipity—the gift of finding valuable or agreeable things not sought for or "the faculty of making fortunate and unexpected discoveries by accident" (dictionary definitions).

The word *serendipity* was coined by Horace Walpole in a letter to his friend Sir Horace Mann in 1754. Walpole was impressed by a fairy tale he had read about the adventures of "The Three Princes of Serendip" (or Serendib, an ancient name for Ceylon, now known as Sri Lanka), who "were always making discoveries, by accidents and sagacity, of things which they were not in quest of. . . ." Walpole used the term to describe some of his own accidental discoveries. The word itself has been rediscovered recently and is being used with increasing frequency. (It does not appear in the 1939 or 1959 editions of a well-known dictionary, but does in the 1974 and later editions and in other current dictionaries.)

Most of the persons who have been blessed by serendipity are not reluctant to admit their good fortune. Far from being defensive about the role that chance played in their discoveries, they are usually eager to describe it. They realize, I believe, that serendipity does not diminish the credit due them for making the discovery. Pasteur, who made breakthroughs in chemistry, microbiology, and medicine, recognized this and expressed it succinctly: "In the fields of observation, chance favors only the prepared mind." More recently, Nobel laureate Paul Flory, upon the occasion of receiving the Perkin Medal, the highest honor given by the American Chemical Society, said:

> Significant inventions are not mere accidents. The erroneous view [that they are] is widely held, and it is one that the scientific and technical community, unfortunately, has done little to dispel. Happenstance usually plays a part, to be sure, but there is much more to invention than the popular notion of a bolt out of the blue. Knowledge in depth and in breadth are virtual prerequisites. Unless the mind is thoroughly charged beforehand, the proverbial spark of genius, if it should manifest itself, probably will find nothing to ignite.

These statements by Pasteur and Flory show that they understood what Walpole meant when he described serendipity as discoveries made "by accident and sagacity."

I have coined the term *pseudoserendipity* to describe accidental discoveries of ways to achieve an end sought for, in contrast to the meaning of (true) *serendipity*, which describes accidental discoveries of things not sought for.

For example, Charles Goodyear discovered the vulcanization process for rubber when he accidentally dropped a piece of rubber mixed with sulfur onto a hot stove. For many years Goodyear had been obsessed with finding a way to make rubber useful. Because it was an accident that led to the successful process so diligently sought for, I call this a pseudoserendipitous discovery. In contrast, George deMestral had no intention of inventing a fastener (Velcro) when he looked to see why some burs stuck tightly to his clothing.

I believe many of the fortuitous accidents that have resulted in the discoveries described in this book should be called pseudoserendipitous, which does not necessarily make them less important. In some examples I point out the distinction between the two classifications; in other instances you may decide whether the discovery was serendipitous or pseudoserendipitous. In either case it is a discovery worthy of our interest and admiration.

Our stories begin with one of the first recorded examples of pseudoserendipity. In the third century B.C., Archimedes was seeking a way to detect the presence of base metal in a golden crown made for his king. He found the answer in an accidental observation he made in the public baths of Syracuse, which caused him to dash naked from the baths shouting "Eureka!"

Such happy accidents have happened to thousands of individuals, although probably most have not reacted as dramatically as Archimedes. We have all profited when discoveries have come out of these accidents. In this book I chronicle some of these discoveries—some earth-shaking, some almost trivial—from early times to the present. As a rule, I have attempted to describe the discoveries in terms the general reader will understand. Some, however, require technical or scientific terminology, or else scientists would find these stories as meaningless as the nonscientist would find some of the others if they were described in precise technical terms. I hope, also, that teachers and students at all levels—from elementary school through university—find these stories useful to enliven lectures and discussions. In some instances, therefore, explanations or additional data are given in formulas and figures. For those who wish further information or background, references are provided in the appendix. Also, a technical supplement containing additional chemical formulas and equations is available on request from the publisher.

For each of the examples described here, there must be hundreds of others of which I am unaware. I hope that this book will be a stimulus for you, if you have encountered serendipity in your own experience or if you know of others who have, to bring these stories to my attention so that I can include them in future editions of the book.

There should be future editions, because serendipity is ongoing—it happens every day!

Acknowledgments

▼ I am indebted especially to Sir Derek H. R. Barton for introducing a lecture at the University of Texas by saying that most of the important discoveries in organic chemistry have been made by accident. He modestly described some of his own brilliant work in terms of "conceptions, misconceptions, and accidents." (You can find some examples of these in Chapter 35.)

At the time I was ripe for encouragement to work on an idea that had long interested me. My interest began with a professional encounter of my own with serendipity (described at the end of Chapter 17). It was also aroused partly by a gift from a student many years ago of a small book titled *Accidental Scientific Discoveries.* This student, whose name escapes me now—if I ever knew it—was kind enough to say that although he or she enjoyed my lectures, this little book might furnish me with "interesting tidbits" for the future. One of my hopes for *this* book is the same—that teachers at all levels may find it a useful resource.

Many others have been helpful in providing me with examples of serendipity that they themselves had encountered, or that they knew of. Among these are my colleagues at the University of Texas at Austin,

Professors Allen Bard, R. Malcolm Brown, Jr., Joseph Carter, Marye Anne Fox, and G. Barrie Kitto, and Dr. Michael S. Brown of the University of Texas Health Science Center at Dallas. Our chemistry librarian, Christine Johnson, was enthusiastic and efficient in obtaining articles not readily at hand. My thanks go also to John Szilagyi, local representative of John Wiley and Sons, Inc., for being interested in serendipity and facilitating the original contact with my publisher; to Dr. Carol Jones of IBM, Austin; to Carl Hesler, Jr., of Austin; to Dr. E. Schoenwaldt, formerly at Merck and Co. and now in Austin; to Professor Prescott Williams of Austin Presbyterian Theological Seminary; to Professor A. T. Balaban of the Bucharest Polytechnic Institute; to Dr. George F. Bass, Abell Professor of Nautical Archaeology at Texas A & M University; to Professor Charles C. Price, my good friend and respected mentor at the University of Illinois; to Professor James G. Traynham of Louisiana State University; to Professor Kenneth Shea of the University of California at Irvine; to Professor A. G. M. Barrett of Northwestern University; to Professor Robert G. Bergman of the University of California at Berkeley; to Professor D. Seebach of the E. T. H., Zürich; to Dr. J. J. (Kim) Wright of the Bristol-Meyers Co.; to Rich Sanders and Judy Borowski of the 3M Company; to P. J. Hannan, formerly at the U.S. Naval Research Laboratory; and to Dr. C. C. Cheng, my valued early graduate student, now a research director at the University of Kansas Cancer Center.

I wish to express my special appreciation to Dr. Roy Plunkett of the Du Pont Company and to Dr. Daniel W. Fox of the General Electric Company for sharing their experiences with me in person and by letter (see Chapters 27 and 32).

I apologize to some of those mentioned above whose stories do not appear in the book. Some of these stories I was unable to describe satisfactorily in nontechnical language—for this book is intended primarily for the general reader—and some I had to give up for lack of space.

I also wish to express my gratitude to those who helped to provide illustrations: to Thelma McCarthy of the National Foundation for the History of Chemistry in Philadelphia; to Ann E. Kottner of the Niels Bohr Library of the American Institute of Physics in New York; to Justin M. Carisio, Jr., Terrence Cressy, and Dianne Currie of the E. I. Du Pont de Nemours Co.; to R. Smelstor and Richard G. Kuhl of Velcro USA, Inc.; and to Joann Jedrusiak of the Squibb Corporation.

Finally, I wish to thank David Sobel for being an extremely capable and understanding editor, with whom it was a pleasure to work on this project.

Contents

TRADEMARKS AND PATENTS

Aralen: patent held by Winthrop Laboratories, a division of Sterling Drug, Inc.

Cyclosporine: patent held by Sandoz AG.

Dacron is a trademark of DuPont, Inc.

Ivory Soap is a trademark of Procter and Gamble.

Librium is a trademark of Hoffman-LaRoche and Company, AG.

Minoxydil: patent held by Upjohn Company.

Mylar is a trademark of DuPont, Inc.

Novocain is a registered trademark of Winthrop Laboratories, a division of Sterling Drug, Inc.

NutraSweet is a trademark of NutraSweet Co.

Orlon is a trademark of DuPont, Inc.

Post-it is a trademark of 3M Co.

Prontosil: patent held by I.G. Farbenindustrie.

Scotchgard is a trademark of 3M Co.

Silver Stone is a trademark of DuPont, Inc.

Silver Stone Supra is a trademark of DuPont, Inc.

Teflon is a trademark of DuPont, Inc.

Terylene is a registered trademark of Imperial Chemical Industries, Ltd.

Valium is a trademark of Hoffman-LaRoche and Company, AG.

Velcro is a trademark of Velcro USA, Inc.

Wheaties is a registered trademark of General Mills.

Xylocaine is a registered trademark of AB Astra.

1 ▼

Archimedes—the First Streaker

▼ Archimedes, the Greek mathematician, lived in Syracuse in the third century B.C. He is known for contributions such as the invention of the lever; the "Archimedes screw" (which is still used in Egypt to raise the waters of the Nile for irrigation); and the law of hydrostatics, sometimes called "the principle of Archimedes." It was he who ran naked from the public baths through the streets of Syracuse shouting "Eureka, eureka!" "I found it".

What had Archimedes found? What so excited him that he forgot to put on his clothes before dashing home? To answer this question we need to learn what Archimedes had on his mind as he stepped into the baths that day. Hiero, the king of Syracuse and a close friend and perhaps even a relative of Archimedes, had commissioned a goldsmith to make a crown for him from pure gold. Upon receiving the finished crown, the king had doubts about whether the goldsmith had put all the gold into it. Couldn't the goldsmith have substituted a less valuable metal, silver or copper, for some of the gold and kept the gold that was not used?

▼ ***Archimedes discovers how to measure the volume of an irregular object, such as a crown.***

It was known how to mix gold with silver and copper. These mixtures, or alloys, retain the rich color of gold even when significant amounts of the other metals are incorporated. Pure gold is called 24-carat gold. The alloy 14-carat gold is 58% gold and 48% other metals; it is commonly used for jewelry and looks almost exactly like pure gold.

King Hiero called in his friend Archimedes and presented the famous mathematician with the job of finding out whether the crown was indeed pure gold and contained all of the precious metal the king had given to the goldsmith. Chemical analysis was not nearly so far advanced in the third century B.C. as was mathematics, and Archimedes was, after all, a very clever mathematician and engineer.

Archimedes had previously worked out mathematical formulas for the volumes of regular solids such as spheres and cylinders. He realized that if he could determine the volume of Hiero's crown, he would be able to tell whether the crown was made of pure gold or of a mixture of gold with other metals.

When he saw water run over the top of the tub as he stepped into the water, he realized that the volume of the overflow water was exactly equal to the bulk of the part of his body that he had placed in the water. Now he saw a way to calculate the volume of any irregular solid object, whether it was his foot or a crown. So if he put the crown into a container filled with water, he could measure the volume of the water that overflowed. This would be equal to the volume of the crown.

Suppose Hiero had given the goldsmith a cube of pure gold that weighed exactly 5 pounds. The edges of such a cube would measure 4.9 cm (cm = centimeter; one centimeter = 0.394 inch) and the volume of the cube would be 118 cubic cm. If the goldsmith made the crown with *all* of this gold and *no other metal,* the crown would weigh 5 pounds, and its volume would be the same as that of the original cube, 118 cubic cm, although in a different shape. If the goldsmith made the crown with only *half* of the gold and substituted *an equal weight* of silver, for example, for the other 2.5 pounds, the alloy crown would weigh 5 pounds, but its volume would be different.

If one could measure the volume of the crown, it would be found to be more than 118 cubic cm because silver is only about half as dense as gold. Density is a measure of the weight per unit volume of a substance. Gold has a density greater than that of any other common metal; its density is 19.3 grams per cubic cm; the density of silver is 10.5 grams per cubic cm, and that of copper is even less, 8.9 grams per cubic cm. A 5-pound crown made of 50% gold and 50% silver would have a volume of 167 cubic cm.

After Archimedes made his accidental discovery at the public baths, it was a simple matter to measure the volume of Hiero's new crown by placing it in water and measuring the volume of the displaced water. When the king found out that the volume was considerably greater than it should have been for a crown made of pure gold, the dishonest goldsmith received swift justice in the form of execution. What was a fortuitous discovery for Archimedes (serendipity!) was not so fortuitous for the goldsmith.

So this serendipitous discovery of a way to measure the volume of any solid object was the cause of the excitement that led Archimedes to dash out of the bath unaware that he had left his clothes behind.

2

Columbus Discovers a New World

Everyone knows that Columbus set out to find a new route to the Orient by sailing *west* rather than *east* and found not the Orient, but a New World, the Americas. This discovery was indeed accidental, even fortuitous in some ways, but was it serendipitous?

There are several elements of accident in the saga of Columbus that are not so well known as his famous voyage. Christopher Columbus was born in Genoa about 1446. He was educated in Pavia in mathematics and natural science, including nautical astronomy. According to a biography by his son Fernando, Columbus first went to sea at about age 15 and may have visited England, Ireland, and Iceland, as well as Greece, Portugal, and Spain. In Portugal he met and married the daughter of a captain in the service of Henry the Navigator, who first explored the Atlantic west of Europe and Africa. From studying the charts of his father-in-law and those of other mariners, he developed a compelling desire to find a new route to the riches of the Orient by sailing west. The modern explorer Thor Heyerdahl thinks that Columbus also knew of letters written four centuries earlier to the Vatican by Norse priests in Greenland describing lands found by sailing farther west.

Columbus realized that an expedition as ambitious as he proposed required royal patronage. He vigorously pursued this support for the ruling monarchs in Portugal, Spain, France, and England. Henry the VII of England turned down the requests, but Isabella and Ferdinand of Spain did not. As a result, much of the New World went to the Spanish rather than to the English.

Another element of accident in the discovery of the New World by Columbus was his incorrect assessment of the size of the world. Although he correctly believed the world to be a sphere, he underestimated its size and he thought that the Asiatic continent was larger and closer to Spain than it was. He based his estimates on the best globe available then. However, the designer of the globe, Martin Behaim, used Ptolemy's circumference for the earth, which was 25% too small. (Behaim's globe still exists; it is now in a museum in Nuremberg, Germany.) After sailing approximately 3,000 miles, Columbus thought that he had somehow missed Japan and that the islands he found belonged to the East Indies (hence his naming the natives "Indians") south of Japan, although these lands were actually many thousands of miles to the west.

But was Columbus's discovery of the New World an example of serendipity? I have emphasized the importance of sagacity in turning some event into a serendipitous discovery. Although Columbus was a brave explorer, he was not sage enough to recognize the significance of his discovery. He died believing he had found new areas of the Orient rather than a new continent. Nor did Columbus profit from his discoveries in the way he had hoped. He received praise and temporary acclaim from his Spanish patrons; but when the goal of magnificent riches from the Orient for himself and his patrons did not materialize, his glory was short-lived and he died a disappointed man. He had hoped to find a short route to the Orient and its riches and was unable to take advantage of what he did find.

The good fortune that perhaps allows us to justify calling the discovery by Columbus "serendipitous" is the development of civilization in the New World, which came long after the historic accidental discovery by the explorer.

3 ▼

A SICK INDIAN Discovers Quinine

▼ The origin of quinine is so clouded that it is difficult to separate legend from fact. According to a story widely accepted in Europe, the wife of the Viceroy of Peru, known as the Countess of Chinchon, was cured of malaria by taking an extract from the bark of a Peruvian tree; she was so impressed by her cure that she carried some of this bark back to Spain in 1638 and thus introduced the use of quinine to Europe. On the basis of this account the Swedish botanist Linnaeus in 1742 gave the name "Cinchona" to the genus of trees from which the medicinal bark was obtained. However, there were two mistakes in this designation. First, although Linnaeus intended to honor the Countess of Chinchon by the name, he misspelled it, leaving out the first *h.* Second, the Countess actually never had malaria and did not carry cinchona bark back to Spain, but died in Cartagena, Colombia, on her way back to Spain.

The first firm record of the use of quinine to cure malaria is that of the Jesuit missionaries in Lima about 1630; hence the name "Jesuit bark" was given to the medicinal bark, about 100 years before Linnaeus. It probably can never be known for sure whether the Jesuits learned of the antimalarial properties of this bark from the Indians. However, an old legend supplies a plausible account of the accidental discovery of the curative properties of the bark of the cinchona tree.

The legend concerns an Indian who, burning with fever, was lost in a high jungle of the Andes. Several species of the cinchona tree (called by the Indians the *quina-quina*) grow on the warm, moist slopes of the Andes mountains from Colombia to Bolivia at elevations above 5,000 feet. As he stumbled through the trees, he found a stagnant pool of water and threw himself to the ground at the edge of the pool to drink the cool water. One taste of the bitter water told him that it was tainted with the bark of the neighboring *quina-quina* trees, which was thought to be poisonous. Caring more for the temporary relief of his burning thirst and fever than for the possible deadly aftereffects, he drank deeply.

To his surprise he did not die; in fact, his fever abated and he was able to find his way back to his native village with renewed strength. He told the story of his miraculous cure to his friends and relatives, and thereafter they used extracts from the bark of the *quina-quina* tree to cure the dreaded fever. The fever was caused by malaria and the chemical the bark contained was quinine. The news of this discovery traveled through the native population and may have reached the Jesuit missionaries in the early seventeenth century. This legend, if true, confirms that even in primitive societies "sagacity" can allow reasoning from an accident to produce a discovery of earth-shaking proportions.

Although the authenticity of this particular legend cannot be ver-

▼ *A fevered South American Indian drinks water from a jungle pool and discovers quinine.*

ified, something like this has happened often. Sometimes the result was fortuitous, as in this case, but often the result was death or injury to the person encountering a potent natural substance for the first time.

▼ POSTSCRIPT

The treatment of malaria with quinine was the first successful use of a chemical compound against an infectious disease. (For the history of the introduction of quinine as a drug into Europe and of transplanting the cinchona tree from South America to southeast Asia, see Chapter 2 in the book by Silverman listed in the Appendix.) The active antimalarial substance in cinchona bark, quinine, was not isolated until 1820 (by the French chemists Pierre Joseph Pelletier and Joseph Bienaimé Caventou), the chemical formula was not known with certainty until 1908, and the laboratory synthesis of quinine was not accomplished until 1944. (For an account of a serendipitous discovery stemming from a naive attempt to synthesize quinine in 1856, see the story of William Perkin and mauve in Chapter 13.)

Malaria is still the most worldwide lethal disease, although in recent years it has been virtually eliminated in *developed* countries by the control of mosquitoes with insecticides. (The malarial parasite is carried from the infected blood of one victim to others by the bites of certain varieties of mosquitoes.) Nevertheless, malaria has killed more people than all the wars throughout recorded history; one cannot overestimate the value of insecticides and drugs that can control this terrible disease. The damage by insecticides to birds and some other animals that has been emphasized should be balanced by consideration of the many thousands of human lives that have been saved by the use of insecticides, especially in view of the newer insecticides that do not cause damage to the environment as did some early ones.

Quinine has been an important factor in world politics. Because of the inaccessibility of the cinchona tree in South America, the trees were cultivated in other areas of the world, notably in the Dutch East Indies. During World War I, Germany was cut off from its supplies of quinine, leading to a major effort in Germany to produce a synthetic substitute. One of the most successful was Atabrine, or quinacrine. In World War II, the United States was fighting in areas that were breeding grounds for malarial mosquitoes (such as North Africa and the jungles of the South Pacific islands) and the Japanese had control of the cinchona plantations. The United States therefore needed to develop effective synthetic antimalarial drugs.

In North Africa, Americans captured Italian soldiers who were carrying pills suspected of being an antimalarial drug; these pills were white, whereas Atabrine was brilliant yellow. The captured pills were brought to the United States and subjected to careful analysis. They were found to be chloroquine, another antimalarial drug developed in the same German laboratory as Atabrine and covered by the same German patent. Pharmacological tests on chloroquine disclosed that it was ten times more effective than quinacrine at the same dosage, had fewer side effects, and it was white.

The U.S. forces in the Pacific were at this time using quinacrine, which was supplied as Atabrine by the Winthrop company, an American affiliate of the German company where the drug had been developed. However, they were not taking their quinacrine pills regularly. The famous Japanese radio propagandist "Tokyo Rose" had convinced them that quinacrine would not only turn their skin yellow (correct), but that it would also make them impotent (incorrect). As a result, they were not taking the pills, and when U.S. troops landed on New Guinea, within two weeks 95% of them came down with malaria.

We who were working on the government antimalarial program of the Committee on Medical Research at the University of Illinois at that time were told about this to impress us with the importance of our project. We were told that a thousand marines with malaria were worse (in a military sense) than a thousand dead marines, because it took other personnel to care for the sick ones. It was at about this same time that we learned that chloroquine was better than Atabrine on a dosage basis, and its not causing a jaundiced appearance was seen to be another big advantage.

There was a rumor that the Italians were using chloroquine in North Africa rather than Atabrine because German testing had indicated chloroquine was less effective and so the Germans gave it to their Axis partners and saved the Atabrine for themselves. The Germans were good chemists, but not such good pharmacologists, however.

When the U.S. pharmacological tests showed chloroquine to be an excellent drug, our antimalarial program gave it a high priority. As a fresh Ph.D. chemist on the program, I was asked to find a new method of synthesizing chloroquine. Obtaining it in pure form by the German process was a problem, perhaps another reason why the Germans preferred Atabrine.

I still remember vividly the excitement I felt when I saw some beautiful white crystals separate from a boiling solution in a critical experiment. These were crystals of an intermediate compound in the synthesis of chloroquine. Before I had more than one gram of chloroquine, Professor Charles C. Price (my research director) and I were on a train to Buffalo to talk to chemists at a manufacturing plant about scaling up our synthesis for large-scale production in order to supply the drug to the armed forces. The process was so successful that Professor Price and I secured patents (in the name of the U.S. Government) and it was used to prepare several tons of chloroquine before the end of World War II.

At another time our entire antimalarial research group (about a dozen doctoral and postdoctoral students and Professor Price) worked in shifts around the clock to make enough of the drug to supply a hospital for clinical tests in progress. We had received a telephone call from New York late Friday afternoon telling us that patients were being tested with the drug and their supplies were running out. I still shudder at the risks we took with makeshift apparatus to make our quota by Monday.

Chloroquine was only one of thousands of new compounds synthesized and tested during and after World War II. It, and some of the other synthetic antimalarials, were used during the Korean and Vietnam wars. After being used for several years, they were found to be less effective—apparently strains of malarial parasites developed that were resistant to these drugs. Because this type of resistance does not develop against quinine, the natural drug has retained its importance in the battle against malaria.

Finally, one other (and slightly facetious) postscript: Someone has said that the British dominated India politically for so long because of the British habit of the daily gin-and-tonic drink. The tonic was quinine, and this kept the British free of the curse of malaria, while many of their Indian subjects, who did not favor this British beverage, suffered the fevers and debilitation of malaria.

4 ▼

SIR ISAAC NEWTON, the Apple, and the Law of Gravitation

▼ It is a rare person who can abstract a universal law from a commonplace event, such as the fall of an apple.

Sir Isaac Newton was born in Woolsthorpe, in Lincolnshire, England, on Christmas day in 1642. His father died before Isaac was born. His mother remarried when he was three years old and left him in the care of his grandmother, who sent him to school at Grantham, about six miles from Woolsthorpe. His mother was widowed again when Isaac was 14, and she returned to the family home in Woolsthorpe. Because he appeared to be only an average student, she brought her son home to run the farm. Isaac was more interested in mathematics and various mechanical hobbies than in farming, however. Fortunately, his uncle, who was a graduate of Trinity College, Cambridge, recognized Isaac's potential and suggested that he be sent back to school to prepare for university. Isaac entered Cambridge in 1661 at age 18. During the next three years in the university, his genius in mathematics and science seems to have awakened. At the same time, the plague broke out in London, and the

▼ *Sir Isaac Newton, 1642–1727*

university closed in the summer of 1665 to prevent further spread of the disease. Newton had received a bachelor's degree early that year; he returned to Woolsthorpe and spent two quiet years there in study and reflection before returning to Cambridge when it reopened.

By the time of his return to the university it appears certain that he had laid the foundations for his monumental work in optics, mathematics, and the physics of gravitation and motion. However, Newton did not fully put forward the law of gravitation, which had its genesis in the serendipitous observation of the fall of an apple, until he published his prodigious *Principia* in 1687, some 20 years after the incident. (The reason for this delay has been much debated. For a thorough discussion of this debate, see the book by Florian Cajori, listed in the Appendix.)

Several sources report Newton's observing the fall of an apple from a tree and the consequences of that observation: Martin Folkes, President of the Royal Society; Voltaire, who is said to have heard it from Catherine Barton, Newton's niece; John Conduitt, Newton's friend, who later married his niece; and Dr. William Stukeley, a physician and personal friend of Newton.

Although Voltaire's anecdote received the most publicity, a more reliable account appears to be by Stukeley in his *Memoirs of Sir Isaac Newton's Life* (1752). Stukeley visited Sir Isaac when the latter was an old man and described a conversation they had:

> After dinner, the weather being warm, we went into the garden and drank thea [sic], under the shade of some apple trees, only he and myself. Amidst other discourse, he told me he was just in the same situation as when formerly the notion of gravitation came into his mind. It was occasion'd by the fall of an apple, as he sat in a contemplative mood. Why should that apple always descend perpendicularly to the ground, thought he to himself. Why should it not go sideways or upwards, but constantly to the earth's center? Assuredly, the reason is that the earth draws it. There must be a drawing power in matter: and the sum of the drawing power in the matter of the earth must be in the earth's centre, not in any side of the earth. Therefore does this apple fall perpendicularly, or towards the centre. If matter thus draws matter, it must be in proportion of its quantity. Therefore

▼ *Sir Isaac observes an apple falling from his garden tree.*

> the apple draws the earth as well as the earth draws the apple. That there is a power, like that we here call gravity, which extends its self thro' the universe.
>
> And thus by degrees he began to apply this property of gravitation to the motion of the earth and of the heavenly bodys, to consider their distances, their magnitudes and their periodical revolutions; to find out, that this property conjointly with a progressive motion impressed on them at the beginning perfectly solv'd their circular courses; kept the planets from falling upon one another, or dropping all together into one centre; and thus he unfolded the Universe. This was the birth of those amazing discoverys, whereby he built philosophy on a solid foundation, to the astonishment of all Europe.

In *Isaac Newton: A Biography* (1934), L. T. More described the apple incident in a more imaginative way, emphasizing the serendipitous aspect of the event as it impinged on the "prepared mind" of Sir Isaac:

> He had just been graduated from college and had been successful enough to be appointed to a scholarship. As a boy, he had spent his days on the farm, meditating on the childish problems which interested him, and now as he comes back, a man, he takes up again his former life; but his mind is now full of profound ideas, and his meditations are to change the course of all future thought. In the long summer afternoons, he sits in the orchard which still stands near the old gray stone house; on one memorable day, an apple falls with a slight thud at his feet. It was a trifling incident which has been idly noticed thousands of times; but now, like the click of some small switch which starts a great machine in operation, it proved to be the jog which awoke his mind to action. As in a vision, he saw that if the mysterious pull of the earth can act through space as far as the top of a tree, of a mountain, and even to a bird soaring high in the air, or to the clouds, so it might even reach so far as the moon. If such were the case, then the moon would be like a stone thrown horizontally, always falling towards the earth, but never reaching the ground, because its swift motion carried it far beyond the horizon. How simple—but even a Galileo, who had solved the problem of the projectile, did not have sufficient imagination to guess that the moon was only a projectile moving swiftly enough to pass beyond the earth. Nor could Huygens, who formulated the laws of centrifugal force and motion, penetrate the secret. Perhaps even more significant of Newton's genius, was the fact that he not only guessed the law of attraction, but he immediately set himself the task of calculating what would be the law of the force which could hold the moon in her orbit.

A statement by Sir David Brewster in his biography, *Memoirs of the Life, Writings, and Discoveries of Sir Isaac Newton,* further supports the authenticity of the apple story: "I saw the apple tree in 1814 and brought away a portion of one of its roots. The tree was so much decayed that it was taken down in 1820, and the wood of it carefully preserved by Mr. Turnor of Stoke Rocheford."

Thus serendipity had a part in the birth of the law of gravitation in the mind of the 23-year-old youth who became one of the most renowned scientists the world has known.

5 ▼

THE ELECTRIC BATTERY and Electromagnetism— from a Frog's Leg and a Compass

The Electric Battery

▼ The Italian physiologist Luigi Galvani (1731–1798) is commonly credited with discoveries leading to the first demonstration of electric current. In 1786 he observed that a dissected frog leg twitched as it lay on a table near an electrostatic generator. (A similar observation had actually been made 30 years earlier by Floriano Caldani.) Galvani followed up his observation with studies of what he called "animal electricity." He hung a frog leg from an iron railing by a brass hook and noted that the lower part of the leg contracted when it came in contact with another part of the railing. (Again, he was apparently unaware of a similar observation by Jan Swammerdam in Holland almost a century earlier.)

Galvani's reports aroused the interest of another Italian scientist, physicist Alessandro Volta. Volta thought that the frog leg on the balcony twitched not because of animal electricity, but because of the difference in potential between two dissimilar metals (the brass, which is mainly copper, of the hook and the iron of the railing), which the animal tissue inadvertently connected. To Volta, the frog muscle and nerves

▼ *Luigi Galvani, 1737–1798*

represented an extremely sensitive electroscope, one that permitted detection of a current much weaker than any that had been studied by apparatus available at that time.

Volta proved his theory of differing electrical potentials of dissimilar metals by inventing the first practical battery, which he described in a letter to the Royal Society in London in 1800. Volta's battery used "cells"

▼ *Alessandro Volta, 1745–1827*

composed of two different metals, such as silver and zinc, separated by moistened disks of pasteboard and connected in series. A combination of such ("galvanic") cells made a battery, the power ("voltage") of which depended on the number of such cells connected.

Batteries produced in this way were the first source of a useful electric current. Previously only electrostatic generators were known; they produced discharges of high voltage but could not deliver a continuous current. Even in its early, crude form the Volta battery made possible some important electrochemical discoveries, such as Sir Humphry Davy's discovery of the elemental metals sodium and potassium.

Electromagnetism

By the end of the eighteenth century, Charles Augustin de Coulomb had discovered the inverse-square law of force, Galvani had observed (and explained incorrectly) the electrical effect of dissimilar metals, and Volta had explained it correctly. An essential link between magnetism and electricity was still missing. The Danish physicist Hans Christian Oersted discovered this significant connection in 1820.

Oersted found that an electric current in a wire passing over a compass deflected the magnetic needle. It is not clear from Oersted's writings whether he actually made this discovery during one of his lectures or first demonstrated it on such an occasion. In any case, it led directly to William Sturgeon's invention in 1825 of the practical electromagnet (Sturgeon was an English shoemaker) and improvements by the American physicist Joseph Henry by 1831. The electromagnet has affected our lives greatly by applications ranging from the doorbell and the telegraph to electric motors.

6 ▼

VACCINATION—Edward Jenner, a Milkmaid, and Smallpox

▼ Millions of lives have been saved with penicillin, sulfanilamide, and related bacteriocidal drugs. (For a discussion of these, see Chapter 24.) However, perhaps even more lives have been saved from disease by the preventive action of vaccination—another accidental discovery.

Until the nineteenth century, one of the great scourges of humankind was smallpox. Only the plague and malaria have killed as many people as has smallpox. I have described (in Chapter 3) the control of malaria with quinine and synthetic antimalarials; insecticides are also useful in eliminating the mosquitoes that carry the disease. The plague was finally controlled in the developed areas of the world by means of sanitation, after the disease was known to be spread by fleas on rats.

Edward Jenner is credited with "having presented to the world a vaccine which has saved many millions of people from a horrible death from smallpox and many more millions from frightful disfigurement" according to E. L. Compere in his 1957 article "Research, Serendipity, and Orthopedic Surgery." Dr. Compere writes:

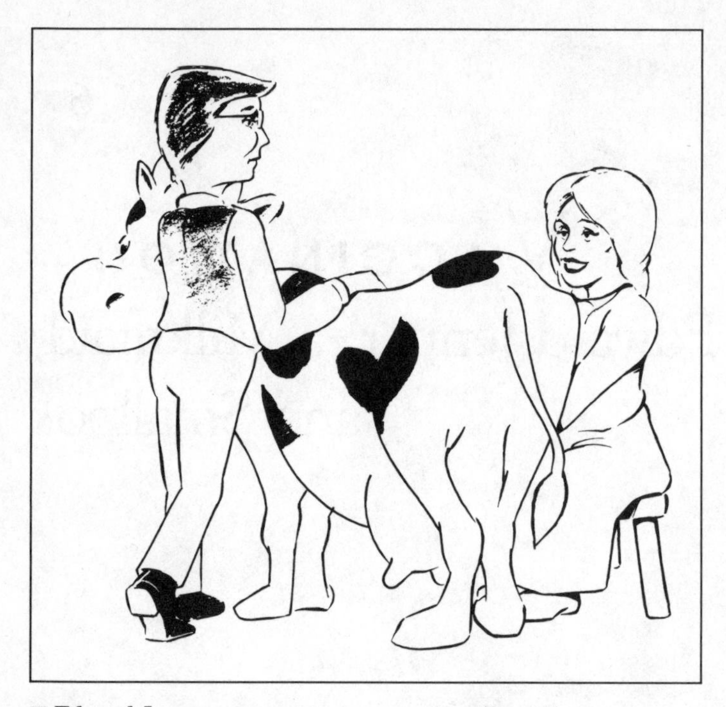

▼ ***Edward Jenner sees cowpox scars on a milkmaid's hands.***

Jenner did not discover his vaccine as a result of long and arduous work in a laboratory. At the age of 19 years he was told by a former milkmaid that she could never have smallpox because she had had cowpox. Jenner recalled this statement when later, as a physician, he realized the futility of trying to treat the disease. He investigated and found that milkmaids almost never had smallpox, even when they helped nurse those who were ill with the disease. The idea of inoculating patients with cowpox in order to prevent them from having the more deadly smallpox occurred to him. This was true serendipity. The fact that cowpox gave immunity to smallpox came to him without effort on his part. He had the good judgment to recognize its value and to make use of it.

Edward Jenner was born in Berkeley, Gloucestershire, in 1749, the son of an English vicar who died when Jenner was six years old. He was raised with the help of an older brother. He received his early schooling in local schools, where he showed an interest in natural history. He began his study of medicine under Daniel Ludlow, a surgeon of Sudbury, near Bristol. During this time, the milkmaid told him about the relationship between cowpox and smallpox.

At age 21 he went to London to study under a famous doctor, John Hunter, and he lived in Hunter's house for two years. He was employed by Sir Joseph Banks to prepare and arrange the zoological specimens that Banks had collected on Captain Cook's first voyage in 1771. He was offered the position of naturalist for Cook's second expedition, but he declined it to pursue his medical practice in Berkeley and, later, Cheltenham. He was interested in ornithology, geology, music, and writing poetry, but by 1792 he had decided to confine his interest mainly to medicine and was granted an M.D. degree by St. Andrews.

Meanwhile, the idea of vaccination must have been maturing in his mind. While in London he had mentioned the connection between cowpox and smallpox to Hunter, who had shown little interest in it. In 1775 Jenner began to investigate the belief of the country people in Gloucestershire about cowpox. By 1780 he had found that there were two different forms of cowpox and only one prevented smallpox. He also determined that the effective form of cowpox protected only when communicated at a particular stage of the disease.

There being few cases of cowpox in his district, he had little opportunity to test his theories. He drew a milkmaid's hand with vesicles (blisters, or "pocks") from cowpox, and took it to London to show to doctors there, but they still did not realize the import of his ideas. However, in May 1796 he inoculated an eight-year-old boy, James Phipps, with matter from the cowpox vesicles on the hands of a milkmaid. In the following July, the boy was carefully inoculated with smallpox matter and, as Jenner had predicted, the boy did not develop smallpox.

One wonders how Jenner persuaded the boy and his parents to take such a chance. Perhaps there was an outbreak of smallpox in the area at that time. A possible explanation is suggested in an article on Immunity in the *Encyclopaedia Britannica* (1962 ed., Vol. 12, p. 116): "Before the discovery of smallpox vaccine in 1796 people were immunized against smallpox by injecting them with material taken from the skin eruption of persons ill with the disease. Some individuals inoculated in this way developed smallpox but, so great was the fear of this disease, many people preferred to take the smaller chance of dying from the inoculated disease rather than develop the more often fatal natural one."

The favorable result with Phipps was extremely encouraging to Jenner, but he waited for a second experiment before announcing his success. This came two years later, because cowpox temporarily disappeared in Gloucestershire.

After the second successful inoculation with cowpox and ensuing protection from smallpox, Jenner prepared a pamphlet to announce his discovery, but he decided to go to London first and repeat the procedure there. However, in London during three months he found no one with the confidence to submit to an inoculation. Just after he returned home, however, Henry Cline, an eminent physician at St. Thomas's hospital in London, carried out several successful inoculations, and he informed the medical profession there of the efficacy of the cowpox protection against smallpox.

Acceptance of Jenner's vaccination procedure was still delayed, however, by two different challenges. A distinguished surgeon, J. Ingenhousz, criticized it severely and prejudiced others against it for some time. By contrast, a rash doctor, George Pearson, sought to claim credit for the vaccination without adequate knowledge or experience and supplied contaminated inoculation material that caused severe eruptions resembling smallpox. Jenner proved that Pearson's vaccine was contaminated, and the news of the success of the pure cowpox material soon spread all over the world.

A measure of the honors ultimately showered on Jenner is indicated by the following list: a Royal Jennerian Society for the proper spread of vaccination in London was established in 1803; Oxford University conferred an honorary M.D. degree on Jenner in 1813; the anniversary of the first successful vaccination (of the boy James Phipps) was for many years celebrated as a feast in Germany; the Chancellor of the Exchequer in England made a grant of 20,000 pounds to Jenner; India raised a subscription of 7,383 pounds for him; statues of Jenner were erected in Gloucester and in London; and there was even one report that Napoleon himself ordered two English prisoners of war released when he was told Jenner interceded for them, exclaiming, "Ah, we can refuse nothing to that name."

▼ POSTSCRIPT

Jenner did not use the term *vaccination* but used inoculation, or "variolae vaccinae," instead. The Latin words meant, literally, "small pocks of the cow." For almost a century, Jennerian inoculation of cowpox was the only immunization procedure against any disease.

In 1880 Louis Pasteur developed an immunization of chickens against a form of cholera, which in an epidemic had destroyed 10% of the French fowls. He isolated a bacterium of this disease and by cultivating an attenuated form of it and inoculating the fowls with the culture, rendered them immune to virulent attacks of the disease. This was the same, in principle, as Jenner's inoculation with cowpox: the smallpox virus was attenuated in the cow before it was transmitted to the milkmaid in the form of cowpox.

Turning next to anthrax, a disease of cattle and sheep, in 1881 Pasteur isolated the bacillus. He cultivated it at a temperature above animal body temperature to produce an inoculation material that would induce a milk attack of anthrax in an animal and render the animal immune for a time against a serious attack of the disease. Pasteur suggested the term *vaccination* for the general procedure of prophylactic inoculation in homage, as he put it, "to the merit and to the immense services rendered by one of the greatest of Englishmen, Jenner."

Four years later Pasteur developed a vaccine for the disease called rabies in animals and (sometimes) hydrophobia in humans. Pasteur's pioneering work, based on the serendipitous discovery by Jenner, made immunization a highly practical science and opened the way to an explosion in the management of infectious diseases. No other single contribution, with the possible exception of the development of antibiotics, has had such a profound effect on human health. According to W. R. Clark in *The Experimental Foundations of Modern Immunology* (1986), the "crowning achievement of the immunization process" has been the total eradication of smallpox. In the first half of this century, between two and three million cases were reported annually. The last case of smallpox in the United States was in 1949, and the last case verified in the world was in Somalia in 1977.

7

DISCOVERIES of Chemical Elements

Most of the early discoveries in chemistry were serendipitous or at least pseudoserendipitous, because there was no theory on which to plan investigations.

The forerunners of chemists, the alchemists, sought to turn other things, mainly base metals, into gold. They tried every conceivable action, reaction, and trick to achieve this transmutation. Even though they never succeeded, they did their best to convince their contemporaries that they had.

For the alchemists, the "elements" were fire, air, earth, and water. Now we know that there are just over one hundred of the simplest forms of matter that we call "elements," and these are the building blocks of the universe. Some elements are common and some are rare. The earth's crust (to about 10 miles deep) is made up mainly (99.5%) of 12 elements. Five elements comprise over 91%: in decreasing order, oxygen, silicon, aluminum, iron, and calcium. If we include the oceans and the atmosphere, hydrogen and nitrogen are among the most common elements (water is 11% hydrogen and air is 76% nitrogen).

Historical ages have been named after the major material that tools and instruments were made of in each period: Stone Age, Bronze Age, and Iron Age. Bronze is an alloy (mixture) of copper and tin, and sometimes small amounts of other metals. Copper and tin are widely available. When melted together, they form an alloy that is stronger than either one. Copper and tin and their alloys resist corrosion by air and water, making them useful materials for tools, weapons, cooking vessels, and other objects. Brass is an alloy of copper and zinc and has been known for centuries. Iron and steel (an alloy of iron with carbon and other metals) were obtained with more difficulty and, therefore, more recently.

The most widespread and abundant metal, aluminum, was the last to be used because of the difficulty of obtaining it from its ores. Aluminum does not occur in the free, or "elemental," state because it is too reactive. It combines readily with other elements, in contrast to copper, silver, and gold. The stability of gold and silver allows them to be found in the elemental state and retain their beautiful luster better than most other metals.

Some elements that are less common in the elemental state or that occur only in a combined state have been discovered accidentally, including the most abundant element, oxygen.

Oxygen

The discovery of oxygen is attributed to both Joseph Priestley, an Englishman, and Carl Wilhelm Scheele, a Swede. Scheele discovered oxygen more than a year earlier than Priestley but did not publish his results until after Priestley, in 1774, announced his experiment and described the unusual properties of the new "air," as he called it. So Priestley received more credit for the discovery.

Joseph Priestley was an unusual man. He was born in 1733 in Fieldhead, near Leeds in England. Raised in a staunch Calvinist family, he prepared for the ministry, but his liberal ideas soon caused him to be considered heretical not only by the Church of England but also by the Calvinists. Nevertheless, in 1767 at age 34 Priestley became pastor of a small dissenting congregation in Leeds. During this time he was also librarian and literary companion to the Earl of Shelburne, Secretary of State to William Pitt.

▼ *Joseph Priestley, 1733–1804*

On one of his frequent trips to London, Priestley met Benjamin Franklin, who awakened his interest in science and became a lifelong friend. He began to dabble in chemistry, which soon became his consuming hobby. He was an experimentalist with great powers of observation, but his scientific background was negligible, so his conclusions from his experiments were sometimes incorrect and often odd.

He lived beside a brewery in Leeds and became curious about its operation, especially the gas that floated over the fermenting liquors. He found that this "air," as he called it, extinguished lighted chips of wood that he held near the liquid, and that the mixture of gas and smoke that drifted over the sides of the vat "fell to the ground." From this observation he deduced that the gas (it was carbon dioxide) was heavier than ordinary air. He learned how to prepare this heavy "air" in his home laboratory. He found that water in which it was dissolved has a pleasant tangy taste—as everyone knows who enjoys soda pop and other forms of carbonated water. Priestley was awarded a medal by the Royal Society in 1773 for his invention of soda water.

Experiments with this gas led him to study other gases that he could produce. About this time he was given a large magnifying glass, or "burning lens", 12 inches in diameter, which could be used to focus sunlight to heat substances to high temperatures. One of Priestley's inno-

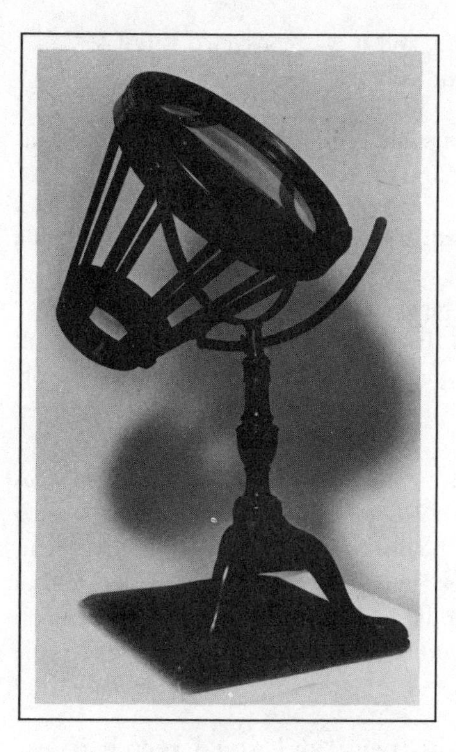

▼ ***The "burning lens" used by Priestley***

vations for studying gases ("airs" as he called them) was an apparatus for collecting them over mercury. He would place substances on the surface of the liquid mercury in a closed glass vessel and heat them with the burning lens; any gases produced would collect above the mercury, which did not dissolve them as water might.

One of the many substances Priestley heated in this way was mercuric oxide, which he called "red calx of mercury." When he heated the red solid, it decomposed and produced a colorless gas above the liquid mercury. Priestley tested this gas with the flame of a candle. Most of the other gases he produced extinguished the candle flame. In *Experiments and Observations on Different Kinds of Air,* which he later published, Priestley described what happened with this "air":

> But what surprised me more than I can well express, was, that a candle burned in this air with a remarkably vigorous flame. . . . I cannot, at this distance of time, recollect what it was that I had in view in making this experiment; but I know I had no expectation of the real issue of it. . . . If, however, I had not happened, for some other purpose, to have had a lighted candle before me, I should probably never have made the trial. . . . A piece

of red-hot wood sparkled in it . . . . and it was consumed very fast . . . I was utterly at a loss how to account for it.

In the introduction to *Experiments and Observations on Different Kinds of Air,* Priestley wrote:

> The contents of this section will furnish striking illustration of the truth of a remark which I have more than once made in my philosophical writings, and which can hardly be too often repeated, as it tends greatly to encourage philosophical investigations; viz. that more is owing to what we call chance, that is, philosophically speaking, to the observation of events arising from unknown causes, than to any proper design, or preconceived theory in this business.
>
> For my own part, I will frankly acknowledge, that, at the commencement of the experiments recited in this section, I was so far from having formed any hypothesis that led to the discoveries I made in pursuing them, that they would have appeared very improbable to me had I been told of them; and when the decisive facts did at length obtrude themselves upon my notice, it was very slowly, and with great hesitation, that I yielded to the evidence of my senses.

Priestley soon discovered that his new "air" would keep a mouse alive twice as long as an equal volume of ordinary air. He also inhaled this new "air" and he reported:

> The feeling of it to my lungs was not sensibly different from that of common air; but I fancied that my breast felt peculiarly light and easy for some time afterwards. Who can tell but that, in time, this pure air may be a fashionable article of luxury. Hitherto only two mice and myself have had the privilege of breathing it.

Two months later, Priestley communicated his results to the eminent French chemist Antoine Laurent Lavoisier, who repeated Priestley's work and made further studies of the new gas. He showed that this gas is the component of (ordinary) air that combines with metals when they are heated in air. He recognized it as a new element and suggested, in 1778, the name *oxygen* (Greek for "acid former," because he thought [incorrectly] that all acids contain oxygen).

Lavoisier first used sensitive scales to measure changes in weight of starting materials and products of chemical reactions. In this way he could show that when heated, mercuric oxide lost weight as oxygen was released, but only to the extent of the weight of the gas released. He proved that the reverse was also true: when heated in air, a metal would increase in weight by an amount corresponding to the amount of oxygen taken from the air. He summarized findings such as these into a statement that became known as the law of conservation of matter: Matter is neither created nor destroyed but is simply changed from one form into another. (We know now that this law must be modified to accommodate the conversion of matter into energy, thanks to Einstein and other modern scientists.)

The discovery of oxygen by Priestley gave Lavoisier the clue to the true explanation of combustion and doomed the "phlogiston" theory, which Priestley continued stubbornly to support until his death. The phlogiston theory dominated chemistry for almost a hundred years, although it was the exact opposite of the correct interpretation of combustion. Combustion is the combination of oxygen with other substances, rather than the combination of the mysterious "phlogiston" with "dephlogistonated air," which was Priestley's definition of his new air. Many scientists consider that modern chemistry began with Lavoisier, his correct theory of combustion, and his law of conservation of matter.

Priestley made two other accidental observations connected with oxygen. Although they were beyond his powers of interpretation, he at least recorded them so carefully that others could profit from them later.

Before the experiment in which he produced the gas by heating mercuric oxide with a magnifying lens, Priestley observed a relationship between combustion, animal respiration, and plant life. He found that air in which a candle had burned until it went out spontaneously was made capable of supporting combustion again and of keeping mice alive *after green plants had grown in the depleted air for some time.* He thus observed the respiration of plants, by which they take in carbon dioxide and produce oxygen, but it was left for others much later to understand this process.

A second observation Priestley called "the most extraordinary of all my unexpected discoveries." It was the observation that a "green matter" that had formed on the walls of the jars used in his experiments liberated a gas when exposed to the sunlight. He recognized the gas as the same as that liberated by heating mercuric oxide, but he could not know that he had first observed the production of oxygen by *photosynthesis.* This process uses energy supplied by the sun, combines carbon dioxide and water, and produces organic matter (Priestley's "green matter") as well as oxygen. Without photosynthesis there would be no life on earth.

▼ ***A caricature by James Gillray showing Priestley calling for the King's head at a Bastille Day celebration in 1791. Although Priestley did not actually attend the celebration, his home was subsequently destroyed by a mob.***

▼ POSTSCRIPT

Priestley came into great personal difficulty because of his liberal religious and political beliefs, which he propounded not only from the pulpit but in writing. He might have survived the accusations of religious heresy, but he also came out openly in support of the French and American revolutions. Because of the strong feelings he expressed about the latter, a mob burned down his church and his home in Birmingham. He moved his family to London, but still suffered persecution for three years until he finally sailed for America in 1794.

Upon landing in New York, he was cordially received by Governor Clinton and other dignitaries. His fame as a theologian, scientist, and liberal had preceded him to the colonies, which had only recently become the United States of America. The Unitarian Church offered him a ministry; the University of Pennsylvania offered him a professorship of chemistry; Thomas Jefferson consulted him in regard to the founding of the University of Virginia; and President George Washington invited him to tea.

He declined the ministry and the professorship, settling for a quiet retirement in Northumberland, a pioneer village in central Pennsylvania. There he spent the last 10 years of his life gardening and experimenting in a laboratory that had been constructed for him. He never became convinced of the error of the phlogiston theory, but he was open-minded enough to admit that he might be wrong. He lived until 1804, long enough to see his friend Jefferson elected President.

Lavoisier's brilliant career was unfortunately cut short by the guillotine in Paris in the same year Priestley went to America. Lavoisier was executed by revolutionaries (he was not only a chemist, but a tax-collector for the ruling aristocracy); Priestley was persecuted by anti-revolutionaries. Neither the French nor the English appreciated these two great scientists at the heights of their careers, but both nations eventually honored them posthumously.

Iodine

Iodine is an element chemically related to chlorine. In an alcoholic solution (called a *tincture*) it is a disinfectant. Bernard Courtois accidentally discovered it.

Courtois was trained as a chemist, but after a few years of study and research at the Polytechnical School in Paris, he decided, in 1804, to follow in his father's footsteps and establish a saltpeter factory near Paris. His business flourished because Napoleon needed saltpeter (potassium nitrate) for making ammunition. The potassium component of the saltpeter was usually produced from wood ash, and the nitrate from decayed vegetable matter.

Seeking a cheaper source of potassium, Courtois found it in the seaweed that washed up on the Atlantic coast of France. From time to time the tanks used in extracting potassium from seaweed ash developed a sludge that had to be cleaned out with acid. One day in 1811 when a stronger than usual acid was used to clean the tanks, a startling sight appeared: violet fumes rose from the tanks and, where they came in contact with cold surfaces, dark, metallic-looking crystals were deposited. Courtois realized that something very unusual had occurred, and he collected some of these strange crystals for further examination.

He found that they would not combine with oxygen but did with hydrogen and phosphorus. With ammonia they formed an explosive compound. Because of the press of business and lack of laboratory facilities, Courtois did not carry the investigation of the new substance further, but turned it over to two friends in the Polytechnical Institute in

Paris, C. Desormes and N. Clement. These chemists described the interesting new material obtained from seaweed in a paper published in December 1813.

At this time, Sir Humphry Davy happened to be in Paris and Clement gave him some of the mysterious substance. When Joseph Louis Gay-Lussac, one of the most eminent French chemists, heard of this, not wishing an Englishman to gain priority on a potentially important discovery, he immediately went to Courtois and obtained a sample of the crystals. After quick and intensive investigations, Gay-Lussac announced that a new element had been found, and he suggested the name *iode*, from the Greek word for violet. Davy confirmed that a new element had been discovered and suggested the name *iodine*, preferring the *-ine* ending to make it conform to the name for its chemical cousin chlorine, which had been named earlier.

To understand the discovery of iodine in seaweed, you need to know that seawater contains other salts besides sodium chloride—among them, although in much smaller amounts, sodium iodide and potassium iodide. The iodide salts become concentrated through biochemical processes in seaweed; when the seaweed burns, the salts become further concentrated. The acid Courtois used to clean out the tanks apparently changed the iodide salts to elemental iodine, which was converted to a violet vapor by the heat of the reaction with the acid; the vapor condensed directly to a crystalline form when it encountered cool surfaces.

Although the discovery of a new element was exciting enough in 1813, an important practical application was not long in coming. In 1820 Jean Francois Coindet, a physician in Geneva, guessed that the new element discovered in seaweed must be the same substance that is present in the ashes of sponges and had been helpful in treating goiter. Analysis of the ashes of sponges showed that they contained iodine, and Coindet then suggested that iodine from seaweed might be used to treat goiter, or hyperthyroidism.

Goiter is a disease caused by lack of iodine in the diet. Biosynthesis of the hormone thyroxine in the thyroid gland requires iodine. Thyroxine controls the rates of many chemical reactions in the body; generally, the more thyroxine, the faster the body works. If the diet lacks iodine, the thyroid gland tries to compensate by enlarging to produce more thyroxine; the enlarged thyroid gland is called a goiter. Goiter has not often been seen in persons living near the sea, because they get sufficient iodine from marine sources. It is common practice now to add small amounts of sodium iodide to ordinary salt (sodium chloride) to prevent goiter from developing in persons living far from the sea.

Helium and the Noble Gases

Helium was discovered not on the earth but on the sun! This discovery, in 1868, was long before space travel, and it was accidental.

Two German scientists, the chemist Robert Wilhelm Bunsen and the physicist Georg Robert Kirchhoff, invented an optical instrument called a spectroscope at the University of Heidelberg in 1859. This instrument could produce a bright-line spectrum (a series of bright lines on a dark background, individually characteristic of each element in terms of the number, color, and spacing of the lines) when the element was heated to incandescence. With this instrument they identified two new elements in the "sodium family" of the periodic table, cesium and rubidium, in 1860 and 1861.

Pierre Janssen, head of the Astrophysical Observatory at Mendon, France, went to India to observe and make photographs of an eclipse of the sun on August 18, 1868. In October, J. Norman Lockyer, who was professor of astronomical physics at the Royal College of Science in London, recorded spectra of the luminous gases surrounding the sun, using a special telescope that allowed this to be done even in the absence of an eclipse. He noted spectral lines that indicated hydrogen to be among the immense volume of gases shooting from the sun. He also saw two yellow lines that were known to be characteristic of sodium, but there was a third yellow line that did not correspond to any known element. He concluded that it might belong to an element present among the gases surrounding the sun but not known on earth. He reported this finding to the Royal Society the same day he observed it, October 20, 1868. Three days later Warren de la Rue reported Lockyer's finding to the French Academy of Sciences.

Meanwhile, Janssen had studied the spectra he recorded in India on August 18, and he found the same new yellow line, which he reported to the French Academy by letter on October 20, only a few minutes after Lockyer's letter had been reported by de la Rue. This posed a problem of priority of credit: Janssen's observation was earlier, but Lockyer's report was earlier. Rather than each insisting on first place, the two astronomers became close friends, and the French Academy struck a commemorative medallion that bore the profiles and names of both men.

Lockyer continued his investigations with the help of Edward Frankland, professor of chemistry at the University of Manchester. He became convinced that the new spectral line belonged to a new element, which he named *helium* after the Greek world *helios* for the sun.

A search then began for evidence of helium on earth, but 23 years passed before any evidence appeared. In 1891 W. H. Hillebrand of the U.S. Geological Survey observed the spectrum of a gas produced by heating a uranium ore; it was mainly nitrogen, but some lines in the spectrum did not belong to nitrogen. When Sir William Ramsey in London read Hillebrand's report, he suspected that the unknown spectral lines might belong to argon, a rare and inert gaseous element that he and Lord Rayleigh had discovered in air a year before. He obtained another type of uranium ore and treated it as Hillebrand had his sample; he did find argon, as he had hoped, but he found an additional yellow spectral line that did not belong to either nitrogen or argon.

He thought at first that the line came from krypton (as it was named later), another inert gas he expected to be associated with argon; but when he sent samples of the gas to Lockyer and to Sir William Crookes for more precise spectrographic measurement, they both confirmed that the wave length of the yellow line was exactly the same as that of helium in the atmosphere of the sun. Ramsay sent simultaneous communications to the British Royal Society and the French Academy of Science announcing the discovery of helium on earth on March 26, 1895. Sir William Ramsay was awarded the Nobel prize in chemistry in 1904 for his discovery of argon, helium (on earth), and the other rare gases. He had found krypton, xenon, and neon between 1895 and 1898, filling up the zero column of the periodic table with the cousins of helium.

The zero column of the periodic table is the position of the rare or "noble" elements in the table in which the elements are arranged according to atomic number and recurring or "periodic" similarities. The Russian chemist Dimitri Mendeleev is usually given the major credit for devising the periodic table of the elements. The elements in the zero column were formerly thought to be incapable of combining with other elements (therefore, zero combining power, or valence); it is now known that even these elements may combine with other elements, but not easily.

In 1904 helium was a rare gas, but in 1905 the situation began to change. Serendipity entered the picture again. A natural gas well near Dexter, Kansas, was capped and the gas piped to a steam generator for use as fuel. To the surprise of all, however, the gas would not burn! When analyzed by scientists from the University of Kansas, it was found to be mainly nitrogen, but more surprising still was that it contained about 2% helium. After this the gases from many other wells in Texas, New Mexico, Utah, and several Canadian provinces were found to contain small amounts of helium. A major source now is in the panhandle of Texas near Amarillo; although the percentage of helium is small (about 1.8%), the volume of gas in this field is so great that it has served as the world's major supplier of helium.

▼ POSTSCRIPT

Beginning in the early years of this century, the Germans developed rigid lighter-than-air ships called zeppelins, after Count Ferdinand von Zeppelin, who designed them. These airships depended on hydrogen for their lifting power. They were used for observation and bombing in World War I and for commercial passenger service in the 1930s. This use ended with the spectacular burning of the Hindenburg when it was landing in Lakehurst, New Jersey, in 1936. It was thought that atmospheric electricity ignited the hydrogen. The 36 lives lost were the first casualties in commercial aviation history.

▼ The Hindenburg disaster—hydrogen was set on fire by lightning. After helium replaced hydrogen for use in aircraft this could not happen.

A similar dirigible was later designed in Germany and modified for helium operation. However, the tense international situation led the United States, which held a monopoly on helium production, to refuse to export helium.

During World War II the United States used nonrigid "blimps" for antisubmarine and shore patrol duties. The lifting power was obtained entirely from helium, which has 93% of the lifting power of hydrogen and none of the fire danger because helium does not burn or support combustion.

The development of modern heavier-than-air aircraft has put an end to the lighter-than-air vehicles for passenger service. However, we can all be reminded of the unusual element that was discovered serendipitously on the sun and later on the earth whenever we see the Goodyear blimp hovering over a football stadium, where it provides a magnificent platform for television cameras.

8

Nitrous Oxide and Ether as Anesthetics

One of the gases that Joseph Priestley discovered and examined with his "pneumatic apparatus" in 1772, two years before he discovered oxygen (described in Chapter 7), was nitrous oxide. It was soon discovered that this gas was nontoxic but produced unusual effects when inhaled: persons turned to antics such as singing, fighting, and laughing. The laughing led to the popular name for nitrous oxide, "laughing gas."

In 1798, Humphry Davy (then 20 years old) was put in charge of the Pneumatic Institute, established by Dr. Thomas Beddoes in Bristol, England, for investigating the medical uses of various gases. Early in the next year Davy discovered that longer inhalation of nitrous oxide produced temporary unconsciousness. Davy tested the gas on himself and reported that he inhaled 16 quarts of the gas during seven minutes and became "absolutely intoxicated." His flamboyant accounts of his experiences drew so much attention to himself and the Institute that Davy was brought to London for a position in the Royal Institution in 1801, where he quickly rose to be a professor. Davy became famous primarily for discovering various chemical elements and establishing their elemental nature. However, he also "discovered" Michael Faraday, the electrical genius, whom Davy made his assistant, and who succeeded Davy as professor in the Royal Institution. Davy was knighted at age 34.

Although Davy suggested that nitrous oxide might be useful in surgical operations, his suggestion was not followed up, and in the early years of the nineteenth century the only use of nitrous oxide was as a source of entertainment. In 1844 in Hartford, Connecticut, a public demonstration was given for amusement, and an accident occurred that led to the discovery of anesthesia in operations.

The demonstrator, a man named Colton, called for volunteers to inhale the gas. A number did, and among them was a young man named Samuel Cooley, who had come with his friend Horace Wells, a dentist. After inhaling the gas, Cooley became violent—he scuffled with others, tripped, and fell. The impact sobered him and he then quietly took his seat in the audience beside Wells. Soon someone saw a pool of blood under Cooley's seat, and it was found to be from a deep cut in his leg. Cooley was unaware of the injury that had caused the wound, and felt no pain until the gas wore off some time later. Being a dentist, Wells saw the significance of this occurrence. Extraction of teeth was very painful in those days. Wells speculated that if the gas could make a person so insensitive as to ignore a severe wound as Cooley had, it might allow painless extraction of teeth.

Wells wasted no time in testing this possibility. He called a dentist friend and asked him to extract his own decayed molar after he had inhaled nitrous oxide. The experimental extraction was carried out in the presence of witnesses, who later testified that they had seen the operation done while Wells was unconscious, and that after Wells regained consciousness he said he had experienced no pain.

Wells next staged a demonstration in the amphitheater of Massachusetts General Hospital in Boston. A patient was found who was willing to inhale the gas and have a tooth extracted while under its influence. Wells, who was apparently nervous and excited, gave the order to have the tooth extracted before the anesthetic had taken full effect, and the patient screamed in pain. Wells was hissed from the amphitheater in disgrace, and he soon had to give up his profession.

In 1846, two years after the dramatic failure of Wells's demonstration, a student and later partner of Wells, William T. G. Morton, decided to try nitrous oxide on his patients. He knew that Wells had actually used it successfully on patients after the ill-fated demonstration. According to Morton, he asked his present partner, Charles T. Jackson, how to obtain nitrous oxide. Jackson told him try ether instead, which, he claimed to have used earlier for anesthesia. Morton did this, with good success; he extracted a tooth from a patient under the influence of ether without pain on September 30, 1846. Shortly afterward he got permission to stage another demonstration in the hospital Wells used. This time Morton removed a tumor from the neck of a patient using ether as the anesthetic. A large audience observed this history-making demonstration and, as a result of the publicity, the stage was set for a great controversy between Wells, Morton, and Jackson. (The correct chemical name for the volatile liquid Morton used is diethyl ether. It is the most common one of a family of related compounds known generically as ethers; when a non-chemist says "ether" he usually means diethyl ether.)

Jackson now had a very different story to tell: He claimed that he had actually persuaded Morton to use ether because he had personally experienced its anesthetic properties much earlier. He described how in the winter of 1841–1842 he was using chlorine in an experiment, when the apparatus containing it broke, and he was almost overcome by the poisonous chlorine. He inhaled ether and ammonia alternatively, finding them very soothing. The next morning he tried more ether, and found it even more soothing to his sore throat, producing loss of all pain and even consciousness. "Reflecting on these phenomena," he said, "the idea flashed into my mind that I had made the discovery I had for so long a time been in quest of—a means of rendering nerves of sensation temporarily insensible, so as to admit of the performance of a surgical operation on an individual without his suffering pain therefrom."

Jackson attempted to discredit not only Morton, but Wells, and even Davy. He incorrectly quoted Davy as having said nitrous oxide was not an anesthetic; he said that Wells had also found that nitrous oxide did not suppress pain and that he himself had confirmed this. The battle between Jackson, Wells, and Morton raged; it continued among their supporters even after Wells committed suicide in 1848. The dispute was finally brought before the Congress of the United States for settlement. At this time another contender for the honor of discovering anesthesia appeared.

The fourth contender was Dr. Crawford W. Long, of Jefferson, Georgia. In the early 1840s there was a fad in the South of holding parties to "get high" on nitrous oxide, much like the demonstration in Hartford, Connecticut, that Wells witnessed. Some of Long's friends asked him for nitrous oxide. Having none, he told them he did have ether, which he knew from his own experience would produce "equally exhilarating effects." They took the ether and, after enjoying the effects upon themselves, they decided it would be fun to try it on the black slave who was serving refreshments. He did not appreciate the idea, however, and, in the scuffling that ensued, his deep breathing caused him to inhale so much ether that he fell to the floor completely unconscious. Dr. Long was quickly called in, the revelers fearing that the man was dead. Examination revealed that he was breathing regularly with a regular pulse, but he could not be roused. When he awoke several hours later, he was normal, except that he had no memory of what had happened.

Long, like Wells, saw the possibility of performing operations on patients while under the effects of an anesthetic, ether in this case, rather than nitrous oxide. Dr. Long's first opportunity to test this theory came on March 30, 1842, when he removed two tumors from the neck of a patient without pain after he administered ether. This was four years before Morton performed an operation on a patient anesthetized with ether in Boston. Long used ether regularly for anesthetizing his patients after this and, being a modest man, he did not publicize his discovery.

When the controversy erupted between Jackson, Morton, and Wells, however, friends of Long brought the claim of his priority before Congress, but Congress declined to make an official decision. Wells received (posthumously) resolutions from the American Dental Association in 1864 and the American Medical Association in 1870 as discoverer of anesthesia in the United States.

Morton and Long continued to use ether in their operations. Enough accidents occurred during ether anesthesia to turn ultraconservative members of the dental and medical professions against its use, and Morton was actually harassed until he was driven out of the profession. He was found unconscious in Central Park in New York in 1868 and died from injuries thought to be the result of attack by an unknown assailant. Jackson received little recognition besides the Order of the Red Eagle of Prussia; he died in a mental institution in 1880. Crawford Long died of natural causes in 1878. The state of Georgia erected a statue of him in the Hall of Fame in Washington, D.C., with the inscription, "Georgia Tribute, Crawford W. Long, discoverer of the use of sulphuric ether as an anesthetic in surgery, on March 30, 1842, at Jefferson, Jackson County, Georgia, U.S.A." (Actually, this ether was technically diethyl ether.)

There have been few discoveries more serendipitous than those of anesthesia by nitrous oxide and ether, few as important to humankind as these, and perhaps no others in which the credit and honor has been as difficult to assign.

▼ POSTSCRIPT

Someone said, "There's nothing new under the sun," and someone else said, "What goes around, comes around." Both statements can be applied to the discovery and use of nitrous oxide. It was used in the early days of the nineteenth century by young people as a "laughing gas," and now modern teen-agers have learned that they can get high on it. It is the propellant for imitation whipped cream, and so it is available at the local supermarket or from restaurant supply dealers. Because of its accessibility, it is being abused by many persons, according to police reports.

▼▼▼▼▼▼▼▼▼▼▼▼▼▼▼▼▼▼▼▼▼▼▼▼▼▼▼▼

9 ▼

WÖHLER'S Synthesis of Urea—Organic Chemistry Begins to Make Sense

▼ Perhaps the most complicated natural substance ever synthesized in the laboratory is vitamin B_{12} (Figure 9.1). In 1972 Robert B. Woodward and Albert Eschenmoser announced its total synthesis. This resulted from collaborative work at Harvard and Zürich by 100 chemists from 19 countries for 11 years. Although the synthesis will never be a practical source of the vitamin, it was a landmark in organic synthesis because new reactions, techniques, and theories were developed during the work.

The first natural substance ever synthesized, urea, was much simpler (Figure 9.2). Friedrich Wöhler synthesized it in his laboratory in Berlin in 1828 accidentally.

Urea was known in 1828 as a typical organic compound. Jons Jacob Berzelius, a renowned Swedish chemist, defined the term *organic* about 1807. It applies to any substance produced by a living *organism*, plant or animal, as contrasted with substances from nonliving, mineral sources, which were defined as inorganic. All known chemical compounds were

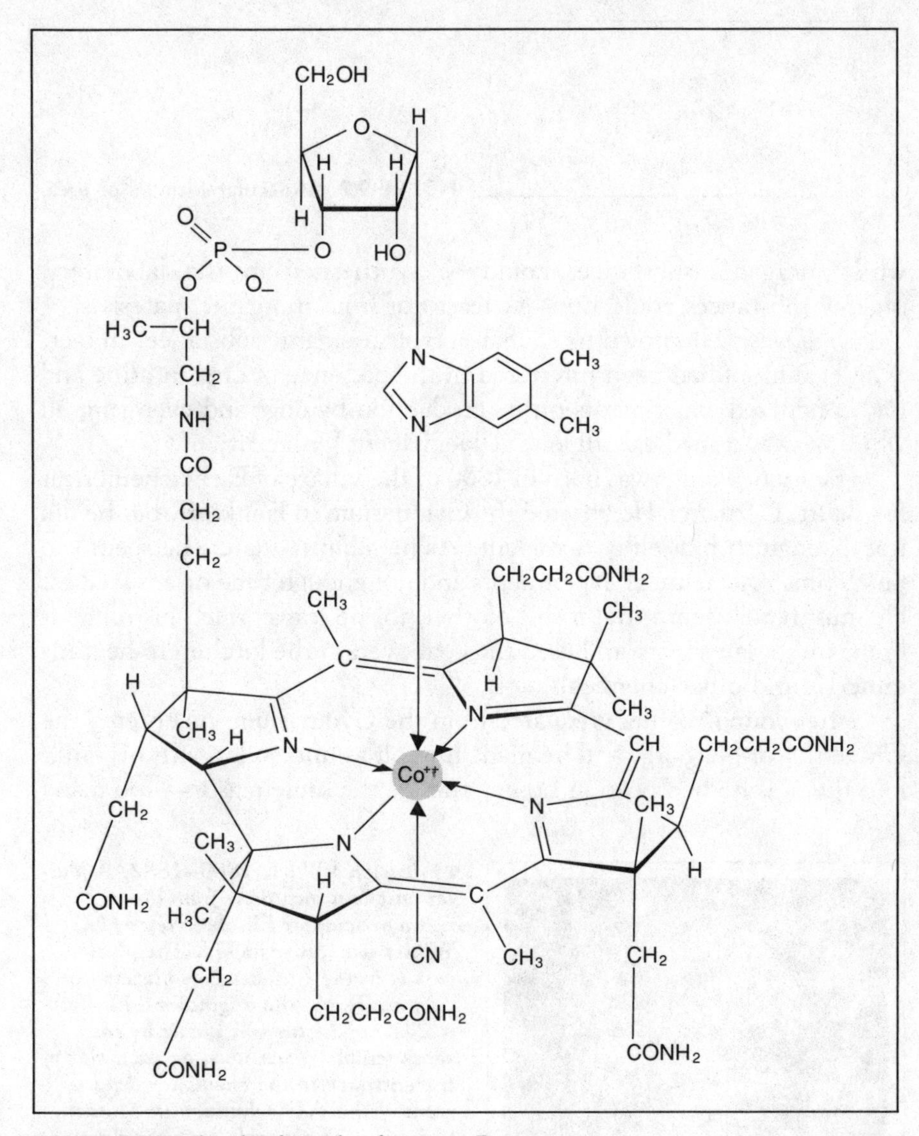

FIGURE 9.1. Molecular formula of vitamin B_{12}

placed in one of these two categories in the early nineteenth century. Inorganic substances, such as the common metallic elements and the compounds of these found in minerals, were much simpler than organic substances, such as sugar, starch, and animal fat. The latter were thought to have a "vital force" that could be transmitted from one plant or animal to another. The theory of vitalism included the premise that

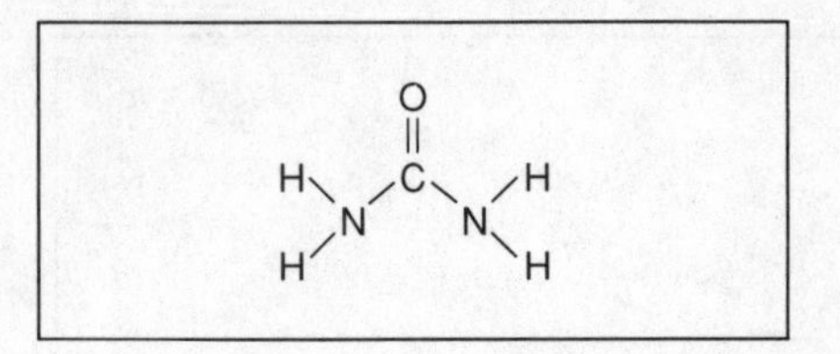

FIGURE 9.2. Molecular formula of urea

while inorganic substances could be synthesized in the laboratory, organic substances could not—at least not from inorganic materials.

Urea was well known in 1828 as a typical organic substance. In fact, Wöhler himself had been interested in the presence of urea in urine and had conducted experiments on its production by dogs and even himself while he was a medical student at Heidelberg University.

Friedrich Wöhler was born in 1800 in the village of Eschersheim near Frankfurt, Germany. He entered the Gymnasium in Frankfurt, but he did not distinguish himself as a student. As he admitted later, he spent too much time on chemical experiments and not enough time on his studies. He must have been a trial to his mother, for he transformed his room at home into a laboratory and used the coal oven in the kitchen to heat his minerals and other chemicals.

After young Wöhler graduated from the Gymnasium, he entered the University of Marburg, and he made himself as unpopular with his landlord there as he had been at home, and for the same reason—too many

▼*Friedrich Wöhler, 1800–1882. A plaster cast of a bust done from life by the German sculptor Elisabet Ney when Wöhler was 68 years old. The plaster cast is in the Elisabet Ney museum in Austin, Texas; the original marble bust and one of Justus von Liebig by the same sculptor once stood on each side of the entrance to the chemistry department of the Polytechnikum in Munich, Germany. The marble statues, along with the Polytechnikum, were destroyed by bombing during World War II.*

experiments in his living quarters! After one year at Marburg, he transferred to Heidelberg, where he came under the influence of Leopold Gmelin, a distinguished German chemist of that period. Although he graduated from Heidelberg as a medical doctor, he was advised by Gmelin, who himself had once been a doctor, to give up practical medicine and devote himself to chemistry.

With Gmelin's blessing, Wöhler went to Stockholm to study and work with Berzelius. He spent only a year there, but Berzelius and he formed a lifelong friendship. He went back to Heidelberg for a short time and then accepted a teaching appointment in Berlin at a technical school. Although this was not a major university position, in fact more like a night school teaching job in the big city, he did have a laboratory of his own and he made good use of it. He was the first ever to produce aluminum in the free metallic state in 1827. The method he used was not practical, however, and it was not until an American student at Oberlin College in Ohio discovered an electrolytic procedure nearly 60 years later that aluminum could be produced industrially. (See the postscript at the end of this chapter.)

In this Berlin laboratory in 1828 Wöhler performed the experiment that assured his fame in the history of organic chemistry. He sought to prepare pure ammonium cyanate, which was thought to have the formula shown in Figure 9.3, from potassium cyanate and ammonium sulfate, two typical inorganic salts. After heating the two salts together, he evaporated the solution containing what he expected to be ammonium cyanate. However, he obtained white crystals that looked exactly like the urea he had so often obtained from dog and human urine! He quickly proved that it was indeed urea. Wöhler described the unexpected result as "a remarkable fact inasmuch as it presents an example of the artificial production of an organic, and so-called animal, substance from inorganic substances."

Another aspect of the "remarkable" result was important to Wöhler, and also to Berzelius, who quickly learned of Wöhler's discovery. Ammonium cyanate and the urea produced from it were "isomers" (from the Greek words for "equal parts")—a term invented by Berzelius to describe

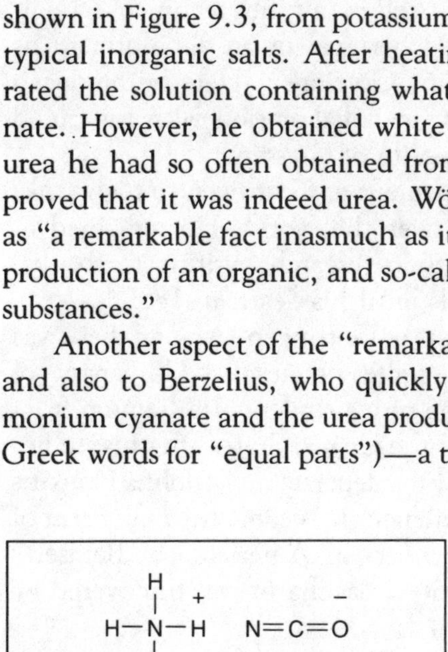

```
      H
      |  +               −
  H — N — H      N = C = O
      |
      H
```

FIGURE 9.3. Molecular formula of ammonium cyanate

compounds composed of the same elements in the same proportions. Ammonium cyanate and urea both contain one carbon atom, one oxygen atom, two nitrogen atoms, and four hydrogen atoms. Wöhler and Berzelius apparently placed more emphasis upon this aspect of the synthesis than upon the impact it might have on the vitalism theory.

However, Wöhler's accidental discovery signaled the beginning of the end of the vitalism theory, which had been holding back the development of the chemistry of carbon compounds. Other chemists of that era pointed out that although usually considered inorganic, the potassium cyanate and ammonium sulfate Wöhler used had been prepared from organic materials such as horns and blood, not from the elements; it was thus not possible to conclude that Wöhler's synthesis invalidated the vitalism theory. Only in 1845 when Hermann Kolbe prepared acetic acid from the elements it contained (carbon, hydrogen, and oxygen), did the purists finally accept the demise of vitalism. The definition of organic chemistry became "the chemistry of carbon compounds" (the currently accepted definition), whether these compounds were obtained from natural sources or from laboratory syntheses.

Wöhler might have devoted himself to organic chemistry for the rest of his career, but he did not—his early interest in minerals persisted and much of his later work lay in inorganic chemistry. After leaving Berlin in 1831, he served for a brief period at Cassell in another technical school, and then in 1836 he finally achieved his goal of an appointment as professor in a major German university, Göttingen. Here he produced some of his best research, carried out in collaboration with his friend Justus Liebig, a professor at the University of Giessen.

As he brought the chemistry department at Göttingen to high rank, Wöhler attracted chemists from all over the world. He remained at Göttingen teaching, training chemists in research, writing textbooks, and editing chemical research journals until his death in 1882.

Wöhler trained about 8,000 students at Göttingen. Among these was Rudolph Fittig, who later became professor at the University of Tübingen, and who in turn had as one of his students Ira Remsen from the United States. Remsen returned to the United States after his studies in Tübingen and developed the chemistry department at Johns Hopkins University to a European level of excellence. It became the major seat of training for future generations of chemists in America. (In Remsen's laboratory at Hopkins the sugar substitute saccharin was discovered by accident, as described in Chapter 22.)

Regardless of his many honors Wöhler's unexpected synthesis of the simple compound urea by heating ammonium cyanate when he was 27 years old stands out as his most important accomplishment. Almost every modern textbook of organic chemistry mentions it. This synthesis of a typical organic compound from a typical inorganic compound marked the end of the vitalism theory and the beginning of organic chemistry on a rational basis.

▼ POSTSCRIPT

The large number of chemists attracted to Göttingen from around the world to study with Wöhler included Frank F. Jewett, who returned to the United States to teach at Oberlin College in Ohio. In the 1880s Jewett often called the attention of his students to the unfortunate fact that although aluminum was the most abundant metal, no one had been able to extract it from its complex ores by a practical process. He told them that his German professor, Friedrich Wöhler, had been the first to prepare the metal, but the procedure he had used was so difficult and expensive that aluminum had remained a museum shelf curiosity for over 50 years.

One of Jewett's Oberlin students was Charles Martin Hall, a local boy. Hall was so impressed with the challenge of the aluminum problem that he decided to find a feasible way to obtain the metal from its ores, and he took this as a project in his senior year at Oberlin. He became convinced that the way to do this was by using electricity. He constructed a makeshift battery and a furnace in the woodshed behind his family home, melted a mineral called cryolite in the furnace, and added the abundant ore of aluminum, bauxite, which he found would dissolve in the molten cryolite. He passed an electric current through the mixture and, to his delight, he saw silvery globules of aluminum accumulate around the negative electrode of his apparatus. As soon as the shiny pellets of metal cooled enough to be held in his hand, he rushed over to show them triumphantly to Professor Jewett. This took place on February 23, 1886, 59 years after Wöhler first produced pure aluminum. The young man who devised the practical process for the production of aluminum was a student of a student of Wöhler!

A few months later a young Frenchman, P. L. T. Heroult, conceived the same electrolytic process, but by this time Hall had applied for a patent on his process, and his patent was given priority. The Aluminum Company of American (ALCOA) grew out of Hall's primitive experiment, using essentially the same electrolytic process. Charles Martin Hall became a rich man and on his death bequeathed a large portion of his fortune to his alma mater, Oberlin College. Anyone who visits the campus now may see a beautiful auditorium dedicated by Hall to his mother, a statue of Hall in the Chemistry Building (where it is now housed in order to protect it from the pranks of students who formerly placed it in less appropriate places about the campus!), and the Hall family home, a block from the campus. The statue and a commemorative plaque on the Hall home are appropriately fashioned of the metal that Hall made available to the world.

10 ▼

DAGUERRE and the Invention of Photography

▼ Have you ever seen a photograph of George Washington? How many photographs of Abraham Lincoln have you seen?

The first successful photographic process was invented by L. J. M. Daguerre in 1835, after Washington's death and before Lincoln became President. Until this invention, people had to rely on artists for the likenesses of famous persons.

Daguerre made his first photograph using a "camera obscura." This was essentially a box with a lens in one end and a ground glass plate where the image was focused. The camera obscura had been invented centuries before; Leonardo da Vinci described one before 1519, and in 1573 E. Danti corrected the inverted image by putting a mirror behind the lens. By the time of Daguerre, people used the camera obscura to trace objects and scenes by placing a sheet of thin paper over the glass plate.

One of the first to attempt to "fix" the image of the camera obscura was another Frenchman, J. N. Niépce. He used a material called asphaltum, or "bitumen of Judea," which became more insoluble in certain solvents after exposure to light. In this way he obtained a more or less permanent image from the camera obscura about 1822. This was probably the world's first photograph, but the product was not satisfactory and the process was impractical. Meanwhile, Daguerre had been experimenting with silver salts, which were known to be sensitive to decomposition by light. When he learned of Niépce's work, he contacted him and they formed a partnership. Niépce died soon after this (in 1833) and Daguerre carried on the work alone, although he made a financial commitment to Niépce's son, Isidore.

Daguerre prepared plates of highly polished silver-plated copper and exposed them to iodine vapor, which produced a thin layer of silver iodide on the surface. Using the camera obscura he exposed these plates, producing a faint image. He tried many ways to intensify this image, but with little success. One day he placed an exposed plate, which had only a faint image and which he intended to clean and use again, in a cupboard containing various chemicals. After several days, Daguerre removed the plate and found, to his amazement, a strong image on its surface!

This was the accident. Now for the discovery that came from the "sagacity" and "prepared mind" of Daguerre: Daguerre concluded that one or more of the chemicals in the cabinet must have intensified the image. So each day he removed a chemical from the cabinet and placed in the cabinet an exposed silver iodide plate. When he had removed all of the chemicals, the image intensification still occurred! Examining the cabinet, he found a few drops of mercury on one of the shelves, spilled from a broken thermometer. He concluded that the vapor of the mercury was responsible for the intensified image, and he soon proved it by experiment. The "Daguerreotype" was the result; thereafter, picture takers developed the latent image by placing an exposed plate over a cup of mercury heated to about 75° Celsius.

Describing his discovery of the mercury process, Daguerre said that he had been experimenting with mercury compounds, from which "it was only a short step to the vapours of metallic mercury, and good fortune led me to take it." (Unfortunately, many Daguerreotype workers suffered severe illness and, some, early death because of the now-known high toxicity of mercury vapor.)

The Daguerreotype was a direct positive photograph. The mercury combined with the elemental silver that had been formed by photochemical decomposition of the silver iodide in the light-struck areas, producing a bright amalgam. In areas not struck by light, the silver iodide was washed away in a later stage of the procedure. The silver-mercury amalgam areas formed a bright image when a dark background was reflected in the mirror of the original silvered plate. (If you view a Daguerreotype with a bright background, as when under a bright sky or light, the image reverses.) The process for removing the unchanged silver iodide was at first a simple washing with a solution of salt (ordinary sodium chloride), but the removal was soon improved when sodium thiosulfate ("hypo") was found to be a better "fixing" agent.

The Daguerreotype was an instant success, mainly because the process caught the attention of several eminent scientists in Paris. One of these was Francois Jean Arago, secretary of the French Academy of Science. Arago announced the invention at a meeting of the Academy on August 19, 1839 and presented a motion to the House of Deputies asking an award for Daguerre and Niépce. The award was granted, probably on account of national pride and fear that the process might fall into foreign hands. These actions brought much publicity to the new photographic process, and it quickly became fashionable in England and America as well as in France. In fact, it was exploited and developed

▼ *An early daguerreotype made by Daguerre himself in 1838, titled "View of the Ile de la Cité with Notre Dame." Daguerre sent a similar view to the Emperor of Austria.*

mainly in the United States, fulfilling the fears of the French government.

Many improvements in photography came quickly after Daguerre's pioneering work, most importantly the negative/positive process. Public fascination with the Daguerreotype gave a tremendous stimulus to photography in the middle of the nineteenth century. The discovery that launched the Daguerreoptype and its inventor into fame was serendipitous.

11

RUBBER—Natural and Unnatural

Vulcanization

In the early sixteenth century, Columbus and other Spanish explorers found South American Indians playing games with a ball formed from the vegetable emulsion called latex exuded by certain trees. One of the names the Indians used for latex was *hevea,* and the principal tree from which they obtained the latex was *Hevea brasiliensis.* Although the Spanish explorers brought some of this "India gum" back to Europe, no good use for it was found until Joseph Priestley, the discoverer of oxygen, showed that it could *rub* out lead pencil marks. Thus it received its present name, *rubber,* from a somewhat trivial, but still useful application.

Europeans found no important use for rubber for more than two centuries because it became soft and sticky at higher temperatures and stiff and brittle at lower temperatures. The Scot Charles Macintosh made one of the few applications of India rubber, overcoming the rubber's stickiness when warm by coating two pieces of cloth with rubber and pressing them together with the rubber in the middle as glue. Macintosh used the double fabric made waterproof in this way to make rain coats. Thus the Macintosh was invented and the name is still used in England for rainwear made of modern fabrics.

Boots and shoes made of rubber or rubber-coated fabric were first made in England and imported into the United States, and were later made in the United States in the 1830s. Americans soon became disgusted with shoes that became stiff in winter and soft and shapeless in summer. At this point Charles Goodyear entered the scene.

Goodyear was born in 1800 in New Haven, Connecticut, the son of an unsuccessful merchant and inventor. The younger Goodyear became fascinated with the possibility of making rubber impervious to temperature changes so that it would be useful in many ways. This fascination became a compulsion that devoured Goodyear's health and the little wealth that he and his family had between 1830 and 1839. During this period Goodyear was in debtor's prison more than once; he became dependent on relatives for food and shelter, but still his obsession persisted. One of his disasters was to sell the government a large order of mailbags that had been impregnated with rubber to make them waterproof, but they turned sticky and shapeless from heat before they left the factory.

After many unsuccessful and unscientific attempts to treat rubber, one of which involved mixing it with sulfur, he accidentally allowed a mixture of rubber and sulfur to touch a hot stove. To his surprise, the rubber did not melt but only charred slightly, as a piece of leather would. Goodyear immediately perceived the significance of this accident. His daughter later said:

> As I was passing in and out of the room, I casually observed the little piece of gum which he was holding near the fire, and I noticed also that he was unusually animated by some discovery which he had made. He nailed the piece of gum outside the kitchen door in the intense cold. In the morning, he brought it in, holding it up exultingly. He had found it perfectly flexible, as it was when he put it out. (Peirce, *Trials of an Inventor: Life and Discoveries of Charles Goodyear,* p. 106)

By further tests, Goodyear determined the optimal temperature and time of heating for stabilizing rubber. He applied for a patent that was granted in 1844 for a process that he termed *vulcanization* after the Roman god of fire, Vulcan.

▼ POSTSCRIPT

When rubber is heated with sulfur, the sulfur atoms link the long chains of rubber polymer molecules, thus stabilizing them and making the rubber matrix as a whole less sensitive to changes in temperature.

Goodyear's accidental discovery of vulcanization of rubber is not serendipity, according to a strict interpretation of Walpole's definition. Instead of finding something accidentally that was not sought, he found a solution accidentally that he had desperately sought. As I mentioned in the Introduction, there are many examples of fortuitous accidents that have resulted in discoveries when they happened to persons who were seeking something that eluded them until the accident occurred. These are not quite what Walpole intended serendipity to mean, but they are related closely enough to be called *pseudoserendipity.*

Goodyear did not live happily even after his discovery of the vulcanization process. He became embroiled in defending his patent and never recovered from his huge debts before he died in 1860, even though Daniel Webster successfully defended him in one patent infringement case. However, the vulcanization process did lead to great activity in the manufacture and use of rubber. By 1858 the value of rubber goods produced reached about $5,000,000. The major rubber manufacturing companies, including the Goodyear Company, were founded in Akron, Ohio, in 1870 and afterward. And this was before the automobile, truck, and airplane, whose tires contain most of the rubber used today.

Synthetic Rubber

The first two commercially successful synthetic rubbers, Neoprene and Thiokol, were both produced by accident. The discovery of Neoprene was pseudoserendipitous and that of Thiokol was serendipitous.

Chemists learned about the molecular structure of rubber by heating it under controlled conditions and identifying the fragments they found. One of these fragments was isoprene, a five-carbon compound containing two double bonds. In 1920 Hermann Staudinger wrote a famous paper that suggested an explanation of the structure of important natural products like rubber, cellulose, and proteins and of certain synthetic substances with similar properties. He proposed that these substances, which seemed mysteriously different from simpler organic compounds, were polymers. (The word comes from two Greek words, *poly* meaning many and *meros* meaning parts.) Polymers are composed of gigantic molecules containing repeated units held together by the same kinds of chemical bonds as in simpler compounds. For example, a molecular formula for rubber was proposed: a large number of isoprene "monomer" (literally "one part") units are assumed to be joined by biosynthesis in the rubber tree to produce the giant polymer molecule of rubber.

After this formula for natural rubber was proposed, many attempts were made to prepare a synthetic rubber that would have the molecular structure and elasticity of the rubber obtained from trees. Isoprene was treated with various catalysts to see whether it could be polymerized to produce something like rubber. These attempts were partially successful, enough to indicate that Staudinger's theory was correct, but subtle factors of molecular structure were not understood until Karl Ziegler discovered stereoregulating catalysts in 1953. (This serendipitous discovery is mentioned in Chapter 26.) It became apparent that natural rubber has an "all-*cis*" arrangement of the monomer units of isoprene; this arrangement can be duplicated in synthetic rubber using the new catalysts, whereas previous catalysts caused a random arrangement of *cis* and *trans* units. Only after this did it become possible to produce synthetic rubber that is almost indistinguishable from the natural substance. The choice between natural and synthetic rubber for tires and other articles today is determined mainly by the price of oil, the starting material for synthetic rubber.

Dr. W. S. Calcott in Du Pont's Jackson Laboratory took note of some work that Father Nieuwland had done at Notre Dame University. Nieuwland was a Catholic priest, the President of Notre Dame, and a chemist. He published results of his research that showed that acetylene, a hydrocarbon with the formula C_2H_2, could be made to add to itself once or twice to produce vinylacetylene and divinylacetylene, molecules with formulas of C_4H_4 and C_6H_6. Calcott thought these dimers and trimers might be similar enough to the unit of which natural rubber is constructed, isoprene, that they could be used to prepare a synthetic rubber. He set some of his Du Pont chemists to work on this job, but they had no success, so he went to Wallace Carothers, the director of Du Pont's Experimental Station, the laboratory where the major polymer research was done.

Carothers was interested. He asked one of his chemists, Arnold Collins, to purify a sample of the crude mixture obtained from acetylene by Nieuwland's procedure. When he did this, Collins separated a small amount of liquid that did not seem to be either vinylacetylene or divinylacetylene and had not been described by Nieuwland. He did not discard it, however, but set it aside on his laboratory bench over the weekend. On Monday he noticed that the liquid had solidified, and when he examined it, he found it seemed rubbery and even bounced when it was dropped on the bench.

One might say that this was no accident, but just what Calcott had hoped for and even expected. However, when they analyzed the rubbery solid, it was found not to be a hydrocarbon polymeric form of acetylene, but contained *chlorine,* which was totally unexpected. Apparently the chlorine had come from hydrochloric acid (HCl), which had been used in Nieuwland's procedure for making the dimer and trimer of acetylene, and it added to vinylacetylene. The product of this addition was named chloroprene, because of its similarity to isoprene. The only difference was the chlorine atom in place of a methyl group (a molecular unit of a carbon atom bound to three hydrogen atoms, that is, CH_3) in the monomer molecule. The spontaneous polymerization of the chloroprene while standing on Collins's lab bench over the weekend produced the rubber-like solid, which DuPont christened Neoprene.

This new synthetic rubber was found to have remarkable resistance to oil, gasoline, and ozone, in contrast to natural rubber. These properties justified its being produced and put on the market by Du Pont in 1930 even though it was much more expensive than natural rubber. It is still useful and valuable; its durability has been proven in such demanding applications as industrial hose, shoe soles, window gasketing, heavy-duty drive belts, and electrical cable jacketing. A recent interesting application is as the bonding material for two-ply leather belts: a sandwich of black and brown leather strips can be bonded permanently without stitching to make a reversible two-colored belt.

In 1924, J. C. Patrick set out to prepare something useful from the large amounts of ethylene and by-product chlorine gas available from industrial processes. These two compounds were known to combine to give ethylene dichloride; Patrick was testing the reaction of various substances with ethylene dichloride, hoping to produce ethylene glycol, which would be a marketable product. One of the things he tested was sodium polysulfide. This did not give the desired liquid glycol, but instead a rubbery, semi-solid material. Patrick recognized immediately the potential value of this unexpected rubbery stuff, and he embarked on an extensive research project that soon led to a patent and the formation of a company to manufacture the new synthetic rubber.

The Thiokol Chemical Company, with Patrick as its president, put Thiokol A on the market in 1929. It was completely different in molecular structure from natural rubber but was elastic. It had one advantage over natural rubber in that, like Neoprene, it was resistant to oils.

Soon, however, a big disadvantage became evident: it smelled terrible!

The Thiokol company and others produced many polysulfide rubbers. Applications for them took advantage of their resistance to petroleum products and good sealing properties, such as sealing windshields of automobiles and lining the fuel tanks in the wings of airplanes. Because Thiokol rubbers could be cured at low temperatures, they were used for a time as a binder and component of solid rocket fuels to propel satellites and space ships into orbit.

In 1982 the Morton Salt Company bought the Thiokol company to form Morton Thiokol, Inc.; both companies had produced specialty chemicals before the merger and they continued to do so afterward. Morton Thiokol received considerable publicity as the major contractor for the construction of the ill-fated Challenger space shuttle. The rubber O-ring whose failure has been blamed for the explosion of the space vehicle was made not of a Thiokol polysulfide synthetic rubber, but of Viton, an elastomer chemically more closely related to Teflon.

12

Pasteur— "Left-Handed" and "Right-Handed" Molecules Make a Difference

Louis Pasteur is better known for his contributions to microbiology than to chemistry. Although he began his professional career as a chemist, he became famous mainly because he first explained how bacteria cause fermentation (his research was invaluable to the production of wine and beer), spoiling of food (the *pasteurization* of milk that he devised is still used), infection of wounds (sterilization during surgery as applied by Lord Lister revolutionized surgical practice), and disease (his investigations of the silkworm disease saved the silk industry in France, and his development of a vaccine for rabies saved thousands of people from death by hydrophobia). However, a contribution he made to chemistry, although less well known and understood, would be enough to assure his fame.

In 1848, when Pasteur was 25 years old and had just received his doctor's degree from the Sorbonne in Paris, he began to study a salt of racemic acid, a substance deposited on wine casks during fermentation. (The name *racemic* comes from the Latin *racemus*, a bunch of grapes.)

▼ *Louis Pasteur in his laboratory*

Another chemist, Eilhard Mitscherlich, had reported that a salt of racemic acid (the sodium ammonium salt made by treating racemic acid with soda and ammonia) was almost identical to a salt of tartaric acid that is also found on wine casks: the salt of tartaric acid was "optically active," whereas the salt of racemic acid was not. (The potassium acid salt of tartaric acid is known as cream of tartar.)

A substance is optically active if it can twist or rotate the plane of polarized light. Ordinary light, according to wave theory, consists of waves vibrating in planes of all directions. Certain crystals can filter light in such a way that only the waves vibrating in one plane pass through. The emergent light is said to be polarized.

It was known in 1848 that certain natural substances (such as quartz crystals, turpentine, and solutions of sugars) could also twist polarized light, but no one understood how they did it. Twisting polarized light was demonstrated and measured in an instrument called a polarimeter. When a substance twists the plane of the polarized light clockwise, it is dextrorotatory—that is, it produces positive rotation. A substance that twists the plane of polarized light counterclockwise is levorotatory—it produces negative rotation.

Pasteur was puzzled: the salts of tartaric acid and racemic acid were said to be identical in chemical composition and crystalline shape, but they had different effects on polarized light. The salt of racemic acid had no effect, whereas the salt of tartaric acid was dextrorotatory. When Pasteur examined the crystals of the salt of racemic acid produced according to Mitscherlich's description, he saw something that Mitscherlich had not. There were, in fact, two kinds of crystals and they were related to one another as a left hand to a right hand.

Using tweezers under a microscope, Pasteur carefully separated the left-handed crystals from the right-handed crystals. When he had enough of each type, he did something that was either acting on a hunch or a flash of genius. He dissolved some of each kind of crystal in water, separately, and placed the solutions in turn in the path of polarized light in a polarimeter. He found that the solution of the left-handed crystals rotated the polarized light to the left and the solution of the right-handed crystals rotated the polarized light to the right. Rene Vallery-Radot in

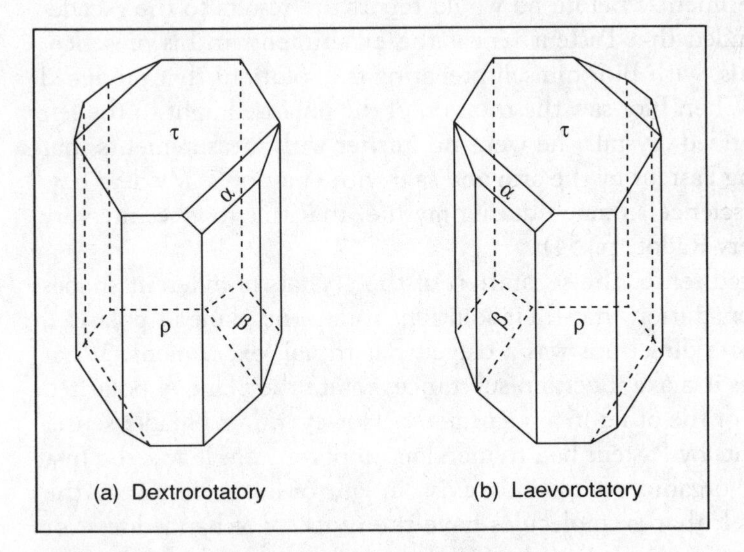

▼ ***Right- and left-handed crystals of a tartaric acid salt***

The Life of Pasteur (1902) reported that the young scientist was so excited by his discovery that he, "not unlike Archimedes," rushed out of the laboratory exclaiming "I have it!" (p. 51)

When he carefully measured the amounts of each kind of crystal as he made the solutions, he found that equal amounts produced exactly the same degree of rotation, *but in opposite directions;* the extent of rotation of the right-handed crystals was the same as that produced by a similar solution of the salt of tartaric acid. Thus Pasteur demonstrated that his right-handed racemic acid salt was actually identical with the dextrorotatory tartaric acid salt, and his left-handed racemic acid salt was a previously unknown mirror-image form of tartaric acid salt. Finally, Pasteur made a mixture of equal amounts of the two kinds of crystals and found, as he expected, that a solution of this mixture was optically inactive. By separating the two kinds of crystals of the salt of racemic acid, Pasteur produced the first and most famous example of what chemists now call a *resolution* of a racemic mixture. (The name of the specific compound, racemic acid, that Pasteur worked with is now used to describe any mixture of mirror-image substances.)

Pasteur's unusual crystallographic experiments became a topic of discussion among scientists in Paris, and the news soon reached Jean Baptiste Biot, the highly respected physicist who had made fundamental discoveries about the rotation of polarized light by crystals. A member of the Academy of Sciences and 74 years old, Biot was skeptical about Pasteur's experiments. Before he would report the results to the Academy, he demanded that Pasteur repeat the experiments in his presence. Pasteur did this, with Biot himself preparing the solutions that produced the crystals. When Biot saw the rotation of the polarized light to the left by the left-handed crystals, he went no further with measurements, but took the young Pasteur by the arm and said with emotion, "My dear boy, I have loved science so much during my life, that this touches my very heart." (Vallery-Radot, p. 54)

In a limited sense, the separation of the crystals of different shapes and the demonstration that their solutions rotate the plane of polarized light in opposite directions was a clever, but trivial, experiment. What difference does it make if certain substances rotate the plane of polarized light one way or the other in a polarimeter? However, in a broader sense, the experiments by Pasteur had tremendous importance. He was the first to show that organic compounds exist in mirror-image forms at the molecular level; that is, molecules have the property of handedness, or *chirality.* This term is based on the Greek word for hand, which is appropriate because hands are the most common examples of mirror-image

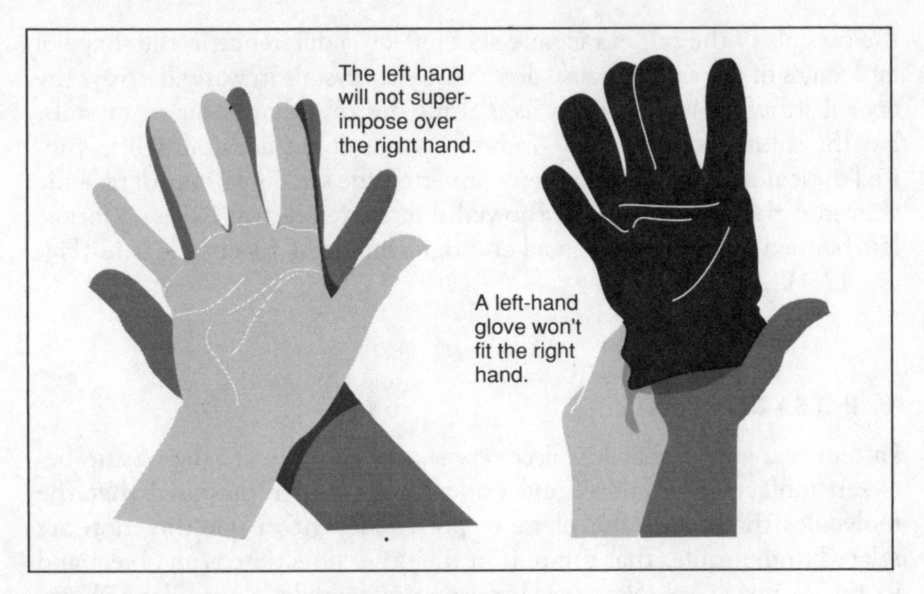

▼ ***Gloves illustrate the characteristic of chiralty***

forms. Not only molecules, but also ordinary objects can be described as chiral or achiral. For example, gloves are chiral, but socks are achiral, because a sock can be worn equally well on a right or left foot.

Before Pasteur's experiments, scientists used the shape of quartz crystals to explain their effect on polarized light, because when the quartz crystals were melted, the optical activity disappeared. However, the optical effects of certain liquids like turpentine and of solutions of sugar defied explanation. Pasteur's work showed that the difference in shape of

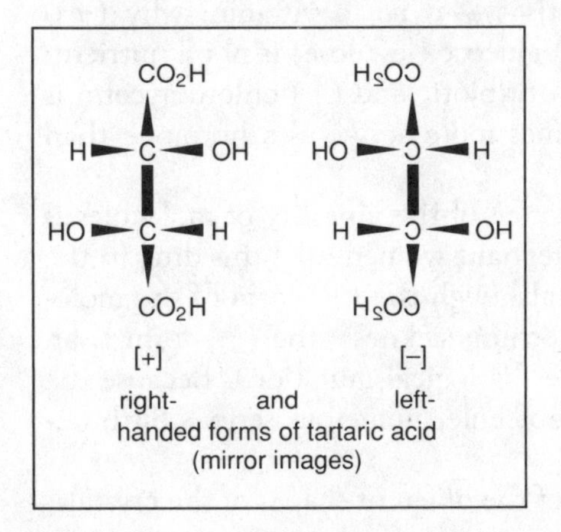

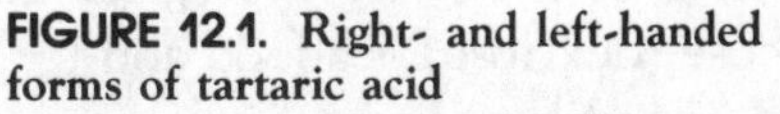
FIGURE 12.1. Right- and left-handed forms of tartaric acid

the crystals of the salt of racemic acid reflects a difference in the shape of molecules of the salt, because dissolving the crystals in water destroys the crystal structure just as surely as melting destroys that of quartz crystals, but the solutions of the two kinds of crystals of Pasteur's salt still exhibited optical activity. Pasteur later converted the separated salts of racemic acid into the acid forms and showed that there are two isomeric forms, left-handed or (−)-tartaric acid and right-handed (+)-tartaric acid (Figure 12.1).

▼ POSTSCRIPT

Pasteur recognized that his discovery suggested a direct relationship between molecular geometry and optical activity. He proposed that the molecules that rotate the plane of polarized light in one direction are related to molecules that rotate it in the other direction as an object and its mirror image are related. However, it remained for two young chemists, Jacobus van't Hoff and Jules LeBel, to explain 25 years later exactly how the atoms could be assembled into these molecular structures. In the interim, Pasteur turned his attention with great success to the biological problems mentioned earlier. But his pioneering work on the resolution of racemic acid led the way for other chemists to explain the relationship of chirality in molecular structure to biological activity, and this is the real significance of Pasteur's work.

Although it is unimportant that left-handed molecules of racemic acid [(−)-tartaric acid] rotate the plane of polarized light counterclockwise, it is tremendously important to understand why a right-handed molecule of vitamin C [(+)-ascorbic acid] *is* a vitamin whereas (−)-ascorbic acid has no biological activity—*is not* a vitamin; why (+)-glucose (dextrose) is a food and (−)-glucose (levulose) is not a nutrient; why (−)-chloromycetin is a potent antibiotic and (+)-chloromycetin is not; why (−)-adrenalin is many times more active as a hormone than (+)-adrenalin.

A tragic example of the importance of the chirality of molecules is the use of the drug Thalidomide. Pregnant women used this drug in the 1950s before chemists realized that although the (+)-form of the molecule is safe and effective against morning sickness, the (−)-form is an active mutagen (an agent that causes biological mutation). Because the drug contained both forms of the molecule, numerous serious birth defects occurred.

Although Pasteur's observation of the different shapes of the crystals, his separation of them, and his deductions about the meaning of their opposite effects on polarized light were indeed acts of genius, serendipity played a large part in his discoveries. Two remarkable elements of chance entered into these discoveries. First, the sodium ammonium salt of racemic acid, which was the one Pasteur examined, is almost the only salt of this acid that crystallizes in mirror-image forms that a person can see are different and can separate mechanically. Second, the crystallization in two forms occurs only at temperatures below 26° C (79° F); above 26° C the crystals that form are all alike and have no optical activity. Pasteur placed the flask containing the solution of the salt on a cool window ledge in his laboratory in Paris and left it there until the next day for the crystallization to occur. Except for fortuitous choice of the right salt of racemic acid and the cool Parisian climate, Pasteur would not have made his remarkable discovery.

Pasteur, like other great beneficiaries of serendipity, recognized the difference between an *accident* and an *accidental discovery,* and he expressed it eloquently in his own language: "Dans les champs de l'observation, le hasard ne favorise que les esprits prepares." ("In the field of observation, chance favors only the prepared mind.") The great American physicist Joseph Henry had the same axiom in mind when he said earlier: "The seeds of great discoveries are constantly floating around us, but they only take root in minds well prepared to receive them."

13 ▼

SYNTHETIC Dyes and Pigments

Perkin and Mauve

▼ In 1856 William Perkin, who had just turned 18, decided on an ambitious project in his home laboratory during his Easter vacation from the Royal College of Chemistry: to prepare quinine artificially. At the college, William was an assistant to the famous German chemist A. W. Hofmann. British Prince Albert had enticed Hofmann to come from Bonn to be the first Director of the new Royal College. In one of his lectures, Hofmann said that it would be highly desirable to prepare quinine artificially, because this drug—the only one effective against malaria—could be obtained only from the bark of the cinchona tree, grown in the East Indies. (For information about the accidental discovery of natural quinine, see Chapter 3.)

Perkin proposed to synthesize quinine from toluidine, derived from coal tar (a cheap by-product of the steel industry), through the "additive and subtractive" method then popular. This method was based on the simple molecular formulas of starting material and desired product. Considering the known formulas of toluidine and quinine, Perkin thought that he could produce quinine by adding a certain number of carbon and hydrogen atoms to toluidine and then adding oxygen atoms to arrive at the proper number and type of atoms in quinine. This was several years before Friedrich Kekule suggested a way in which atoms join to form three-dimensional *structures* for molecules. (For information about Kekule and molecular structures, see Chapter 14.) Just how naive Perkin's plans were becomes apparent when we consider that the structural formula of quinine was not determined until 1908, and its synthesis defied the best chemists until 1944.

Perkin carried out the proposed reactions, first adding three carbons and four hydrogens in the form of an "allyl" group to toluidine and then using the powerful oxidizing agent potassium dichromate. The result was an unpromising reddish-brown sludge. Rather than give up, Perkin decided to try a simpler starting material, aniline. (Actually, the aniline Perkin used contained small amounts of toluidine, which was essential to the formation of the purple dye.) This time the product was an even less promising black solid; but on examining it before throwing it out, Perkin noticed that water or alcohol used to wash it out of the flask turned purple.

Fascinated by this unexpected result, the young chemist tested the purple solutions and found that they would dye cloth. Perkin quickly found a practical way to extract the purple dye from the black mixture. He sent a sample of his synthetic dye to a well-known British dye works for trials on silk and cotton. The assessment was that it looked very promising for silk, but not for cotton. However, the dyers soon found that by pretreating the cotton, they could make the dye practical for cotton.

Thus, a naive attempt to synthesize quinine resulted in the accidental production of the first artificial dye. With the enthusiasm of youth, Perkin decided to patent his dye, build a factory, and go into the dye business. He got no encouragement from his professor, Hofmann, who wanted him to continue in academic studies and research, certain that an attempt to start a dye business was foolish.

Much as I hate to admit it (being a professor), Hofmann was wrong. Fortunately for young Perkin, his father was wealthy and had a great deal of confidence in his brilliant son. With the backing of his father and brother, Perkin got a patent on his dye, built his factory, and solved tremendous problems of upgrading the procedure to industrial scale. Explosions were not uncommon in the early stages of this development; to "control" the process, workmen stood guard with a water hose to spray the cast iron reaction vessel if the contents boiled too furiously!

Nevertheless, his business was a great success. His dye, which was called aniline purple, Tyrian purple, mauve (or mauveine), became extremely popular. The last name was given to the new dye in France, and it became the name most widely used.

Until Perkin's discovery, the only permanent purple or lavendar dyes were extremely expensive. Used since 1600 B.C., the natural dye previously called Tyrian purple could be extracted only from small mollusks in the Mediterranean Sea; these were hard to collect, and 9,000 mollusks were required to produce one gram of dye. Only royalty could afford such a color—hence the association of purple with royalty. Perkin's synthesis of a beautiful, permanent purple dye from coal tar made it available at a price almost anyone could afford. Besides, the old purples were "so fugitive that if a lady put a new violet ribbon on her hat in the morning, it was liable to have a red colour by the evening."

The success of mauve signaled the birth of the synthetic dye industry. However, the Germans, rather than the British, saw its potential and developed it on an enormous scale. Even so, the progress of the new dye industry in England under Perkin's leadership was so rapid that only six years after his first factory began, the Chemical Society asked Perkin to lecture on dyes derived from coal tar.

Perkin's later life was filled with honors: election to the Royal Society, knighthood, the Davy Medal, the Hofmann Medal, the Lavoisier Medal, a gala celebration of the fiftieth anniversary of the founding of the coal tar dye industry attended by some of the world's most distinguished scientists in 1906, the Perkin Medal established by the American Section of the British Society of Chemical Industry as a token of its highest honor for American chemists.

The discovery of mauve by Perkin is a good example of serendipity: he set out to do one thing and, through a fortuitous accident, ended up doing something different but much more important. There was no way that he could have synthesized quinine in 1856, and even if he had, the achievement of this goal would probably not have been as important as the founding of the synthetic dye industry. R. B. Woodward's and William von E. Doering's synthesis of quinine in 1944 took genius, but it was not practical. Even though the United States was fighting a war with Japan in malaria-infested areas of the Pacific and had been cut off from access to quinine, the Woodward-Doering synthesis was never used for the practical production of quinine.

▼ *William H. Perkin in the United States in 1906 to attend a celebration of the fiftieth anniversary of Perkin's discovery of the dye mauve*

▼ POSTSCRIPT

Another important coal tar dye that Perkin developed and manufactured was alizarin, a red dye that had been obtained from the root of the madder plant for centuries. In 1868, Carl Graebe and Carl Liebermann in Germany announced a synthesis of alizarin from anthracene, a component of coal tar. The original synthesis by Graebe and Liebermann was impractical for commercial production, but its announcement aroused Perkin's interest because of his familiarity with anthracene while studying with Hofmann. In less than a year he had worked out a practical commercial synthesis of alizarin starting with coal tar anthracene. By the end of 1869 Perkin's factory had produced a ton of alizarin and by 1871 it was making 220 tons per year.

In 1874, Perkin sold his factory. At age 36 he was wealthy enough to devote the rest of his life to pure research. He bought a new house, but he continued to use as his laboratory the one where he discovered mauve. Here he achieved the synthesis of coumarin, the first perfume from coal tar. He prepared cinnamic acid (related to cinnamon, as the name suggests) by a method so generally useful that it became known as the Perkin reaction. Adolf von Baeyer used a variation of this method in 1882 to prepare a starting material for a famous synthesis of indigo, another milestone in the history of dyestuffs.

The value of coal tar dyes is not limited to coloring fabrics; they are used as staining agents in microbiological research. Researchers discovered the bacilli of tuberculosis and cholera by using such dyes in staining techniques.

The discovery of mauve by Perkin has been credited with starting the tremendous development of organic chemistry in the latter half of the nineteenth century, especially in Germany. The fact that Hofmann went back to Germany had something to do with this explosion of knowledge, but more important was the growth of "aromatic chemistry" in Germany, stimulated by Kekule and his chemical architecture. Many of the new synthetic dyes were aromatic compounds related to benzene.

Graebe and Liebermann and Alizarin

Alizarin, a red dye, has been known since ancient times. Egyptians used it to dye the cloths in which they wrapped mummies. It was obtained from the roots of the madder plant, various species of which grow throughout the world.

In 1868 its chemical composition was unknown, but it became the subject of investigation in the laboratory of Adolf von Baeyer, in Berlin. A few years earlier, Baeyer had embarked on studies of indigo, another natural dye (see the next section). In the work on indigo, Baeyer developed a new procedure for removing oxygen from complicated organic compounds to convert them into simpler and, it was hoped, known compounds. He suggested that this procedure, which consisted of heating the compound with powdered zinc, be applied to alizarin. Two of his young associates, Carl Graebe and Carl Liebermann, did this and obtained a product that they quickly identified as anthracene, a well-known hydrocarbon constituent of coal tar.

They did this just a few years after Kekule proposed his cyclic structure for the benzene molecule, and so they could formulate anthracene as three benzene rings fused together. As I mention in the Kekule story, his proposal of a pictorial molecular theory paved the way for the elucidation of structures of many organic compounds, especially the aromatic ones, such as alizarin.

Graebe and Liebermann then set out to reverse the process—to produce alizarin by *adding* oxygen to anthracene. They based their plans on what are now recognized as misconceptions about the possible chemical reactions, but pseudoserendipity stepped in. Although they used illogical procedures, they soon had a synthetic product identical with natural alizarin! This was the first time anyone had synthesized a natural dye in the laboratory. (Perkin's synthetic mauve was a new substance, similar in color to natural Tyrian purple from mollusks, but not in chemical formula.)

As remarkable as the accomplishment was, the laboratory synthesis was entirely unsuitable for commercial production of alizarin. With the help of Heinrich Caro, a technician with the Badische Anilin- and Soda-Fabrik Co. (BASF), Graebe and Liebermann tried other approaches for a synthesis that would be practical. Initial attempts failed, but Caro eventually discovered, by a completely *unintentional* experiment, an unidentified intermediate that could be converted in good yield to authentic alizarin. This psuedoserendipitous procedure was the same one Perkin discovered independently and almost simultaneously in England. Synthetic alizarin was put on the market both in Germany and in England in 1871 and soon supplanted the natural dye.

A Broken Thermometer and Indigo

Like alizarin, the blue dye indigo was known and used by ancient civilizations. Until the last decade of the nineteenth century, it was obtained only from plants. In India in 1897 nearly two million acres were used for cultivating indigo plants. At about that time a German chemical plant developed a process to synthesize the dye and began to sell the synthetic dye at a lower price than the natural product. An economic upheaval ensued in India and other countries that produced natural indigo. As unbelievable as it may seem, this economic and cultural revolution can be traced to the accidental breaking of a thermometer in a laboratory experiment.

Adolf von Baeyer began his studies of the chemical structure of indigo at Berlin University in 1865. By 1883 he was satisfied that he had deduced the correct structure, and he proved it in the way organic chemists do; that is, by synthesizing the compound with this structure and showing that it was identical with the natural dye in all of its properties. Actually, he devised several syntheses, in one of which he used the Perkin reaction, but none of these could be adapted to commercial production of synthetic dye that could compete in price with natural indigo.

Karl Heumann at BASF developed the first successful commercial synthesis about 1893. This synthesis succeeded because it used as the starting material naphthalene, a component of coal tar and at that time practically a waste product of the steel industry. (Coke is used to convert iron into steel; when coke is prepared from coal by heating, coal tar is driven off and is collected as a viscous, evil-smelling, black liquid that was once thought to be useless. Since the discovery of the synthetic dyes and pigments, it has proven to be a rich source of organic starting materials.)

What accident led to the commercially successful synthesis of indigo? A chemist named Sapper at BASF was heating naphthalene with fuming sulfuric acid and accidentally broke a thermometer; the mercury in it fell into the reaction vessel. Sapper noticed that the reaction did not proceed in the usual way, and he found that the naphthalene had been converted into phthalic anhydride. Further investigation revealed that the sulfuric acid converted the mercury into mercury sulfate, a compound that served as a catalyst for the oxidation of naphthalene to phthalic anhydride, which could readily be converted into indigo.

BASF started selling synthetic indigo in 1897 at a lower price than natural indigo. The process for producing synthetic indigo has changed and improved since 1897, and natural indigo never regained a position on the dye market.

Dandridge and Monastral Blue

Many other examples of serendipity can be found in the history of dyes and pigments. One that occurred much later than those of mauve, alizarin, and indigo is the discovery of a beautiful blue pigment by A. G. Dandridge in 1928. A pigment, as distinguished from a dye, is an opaque insoluble powder that is used to color another material. Pigments are used chiefly in protective and decorative coatings, printing inks, plastics, and rubber.

Dandridge was a chemist at Scottish Dyes Ltd., who operated a plant for the production of phthalimide by passing ammonia into molten phthalic anhydride in an iron vessel. (Phthalimide and phthalic anhydride derive their unique spelling from the fact that they both can be derived from naphthalene.) He noticed some blue crystals on the side and cover of the vessel, and he was sufficiently intrigued to collect some and examine them. Further investigation by Dandridge and his associates proved that the blue crystals had come from a chemical reaction between the iron vessel and its contents, and that other metals such as nickel and copper could take the place of iron in the reaction to produce other pigments.

In 1929, Imperial Chemical Industries, the parent company of Scottish Dyes, Ltd., sent samples of the pigments to Professor R. P. Linstead at Imperial College in London, who undertook an investigation of them because he thought that they "might prove of academic interest." He was certainly correct. Linstead and his colleagues elucidated the chemical structures of the pigments, which were named phthalocyanines, and described them in a series of research papers in 1934. This work, together with X-ray crystal examinations by J. M. Robertson in 1935, led to the structural formulas such as that for a copper-containing pigment. C. J. T. Cronshaw, in giving a general account of the discovery of the phthalocyanines (in *Endeavor,* 1942), commented:

> Although the existence of the phthalocyanines was not predicted, and was perhaps not even predictable, yet now the discovery has been made and the structure of the molecule ascertained, no one can fail to remark the inevitability of the compound. Its right of existence is almost declamatory! It is remarkable how readily at the appropriate temperature of reaction, and in the presence of a metal such as copper, the four integral components almost snap into position. (p. 79)

The formulas of the phthalocyanies are similar to those for the pigmenting units of blood (hemin, in which the central metal atom is iron [Fe], in place of copper [Cu]) and green plants (chlorophyll, in which the central atom is magnesium [Mg]).

Not only were the pigments interesting from an academic viewpoint, but they were also of practical value. Linstead and others took out 26 patents between 1933 and 1942, and many more have since been granted. Replacement of the iron atom in the compound by copper was found to produce an even better blue pigment, called "monastral blue"; it is the best available blue pigment for the three-color process used in color printing. Substituting several chlorine atoms for hydrogen atoms in copper phthalocyanine yields excellent green pigments.

The phthalocyanine pigments have become some of the most valuable coloring materials for printing inks, artist colors, paints, and lacquers. They can also be used to color rayon and acetate. These valuable pigments furnish another good example of a discovery by accident and sagacity—serendipity.

14 ▼

Kekule: Molecular Architecture from Dreams

▼ In the early eighteenth century, the theaters and other public buildings in London were lighted by a gas manufactured from whale oil. When this gas was compressed for distribution in tanks, a volatile aromatic liquid separated. The famous scientist Michael Faraday examined this liquid and determined about 1825 that it contained only carbon and hydrogen, in equal proportions. Later this liquid, which was named benzene, was shown to be a component of the tar distilled from coal in the production of coke. Related aromatic compounds were found in many natural sources.

Benzene posed a major theoretical problem for chemists because of its unusual properties. Most compounds that contained only carbon and hydrogen and in which the ratio of hydrogen atoms to carbon atoms was low behaved differently from benzene. (The ratio was 1:1 in this case, because the molecular formula of benzene was known to be C_6H_6.) They were said to be unsaturated with respect to hydrogen; that is, they would add several molecules of hydrogen, but benzene would not do this. There were other odd things about benzene; no one had been able to suggest a suitable structural formula for it before 1865. The man who did was Friedrich August Kekule.

Kekule was born in Darmstadt, Germany, in 1829. He entered the University of Giessen to study architecture. At Giessen, however, he came under the influence of Justus von Liebig, whose dynamic lectures persuaded him to devote his life to chemistry. From Giessen he went to Paris to study with Jean Baptiste André Dumas and Charles Adolphe Wurtz, then to England, where he associated with the best English chemists. Returning to Germany, he taught first at Heidelberg and then in 1858 he went to Ghent, Belgium, as professor of chemistry. He remained at Ghent until 1865, when he was called to Bonn to take the chair vacated by A. W. Hofmann. He held the post at Bonn until his death in 1896, the same year in which Alfred Nobel died. Three of the first five Nobel prizes in chemistry went to his students: Jacobus van't Hoff (1901), Emil Fischer (1902), and Adolf von Baeyer (1905). Kekule was recognized as one of the greatest teachers of chemistry of the nineteenth century.

In spite of his renown as a teacher, Kekule is best known by chemists for his theories about the molecular structure of organic compounds. Before 1858 organic chemists were, in a sense, working in the dark. They accomplished some remarkable things, but they had no mental picture of what the substances they worked with looked like at the molecular level.

For example, Friedrich Wöhler showed in 1828 that urea was different from ammonium cyanate (see Chapter 9), although both contained carbon, hydrogen, oxygen, and nitrogen in the ratio of 1:4:1:2. They were said to be "isomers," but no one understood the different way in which the same number of the same atoms were connected.

The proposal of a satisfactory structural formula for benzene by Kekule in 1865 was important enough to the scientific community to justify a huge celebration in Berlin's City Hall in 1890, the twenty-fifth anniversary of the announcement of the formula. The development of the synthetic dye industry in Germany and, indeed, the flowering of organic chemistry there in the latter part of the nineteenth century owed a great deal to the structural theories of Kekule and his students and colleagues. At the celebration, the honoree made a speech that was published in the major German chemistry journal. The following quotations are taken from an English translation of the speech published in 1958, the hundredth anniversary of the general theory of structure described by Kekule in his talk.

You are celebrating the jubilee of the benzene theory. I must first of all tell you that for me the benzene theory was only a consequence, and a very obvious consequence of the views that I had formed about the valences of the atoms and of the nature of their binding, the views, therefore, which we now call valence and structural theory. What else could I have done with the unused valences? During my stay in London I resided in Clapham Road. . . . I frequently, however, spent my evenings with my friend Hugo Mueller. . . . We talked of many things but most often of our beloved chemistry. One fine summer evening I was returning by the last bus, riding outside as usual, through the deserted streets of the city. . . . I fell into a reverie, and lo, the atoms were gamboling before my eyes. Whenever, hitherto, these diminutive beings had appeared to me, they had always been in motion; but up to that time I had never been able to discern the nature of their motion. Now, however, I saw how, frequently, two smaller atoms united to form a pair; how a larger one embraced the two smaller ones; how still larger ones kept hold of three or even four of the smaller; whilst the whole kept whirling in a giddy dance. I saw how the larger ones formed a chain, dragging the smaller ones after them but only at the ends of the chain. . . . The cry of the conductor: "Clapham Road," awakened me from my dreaming; but I spent a part of the night in putting on paper at least sketches of these dream forms. This was the origin of the "Structural Theory."

Something similar happened with the benzene theory. During my stay in Ghent, I lived in elegant bachelor quarters in the main thoroughfare. My study, however, faced a narrow side-alley and no daylight penetrated it. . . . I was sitting writing on my textbook, but the work did not progress; my thoughts were elsewhere. I turned my chair to the fire and dozed. Again the atoms were gamboling before my eyes. This time the smaller groups kept modestly in the background. My mental eye, rendered more acute by repeated visions of the kind, could now distinguish larger structures of manifold conformation: long rows sometimes more closely fitted together all twining and twisting in snake-like motion. But look! What was that? One of the snakes had seized hold of its own tail, and the form whirled mockingly before my eyes. As if by a flash of lightning I awoke; and this time also I spent the rest of the night in working out the consequences of the hypothesis. (Benfey, *Journal of Chemical Education,* Vol. 35 1958, p. 21)

Kekule's dreams on the top deck of the bus in London and by the fireside in Ghent led to profound theories of organic molecular structure, which were of enormous value to the development of the science. The

▼ ***Kekule dreams of a snake biting its tail and awakes to propose a cyclic structure for the benzene molecule.***

first dream, in which the atoms "formed a chain," "a larger one embraced the two smaller ones" and "still larger ones kept hold of three or even four of the smaller," led Kekule to propose that certain carbon atoms could link together in chains, with hydrogen atoms and other atoms connected to them. For example, methyl alcohol and ethyl alcohol, whose simple formulas were known to be CH_4O and C_2H_6O, could be represented by the structural formulas in Figures 14.1 and 14.2. Similarly, the structural formula of urea could be written as in Figure 14.3, whereas that of ammonium cyanate would be written as in Figure 14.4.

The second dream, in which the snake was biting its own tail, led Kekule to propose a cyclic structure for benzene, with the six carbon atoms in a ring. As you can see in Figures 14.1 through 14.4, carbon is joined to the other atoms by four lines, in some instances with two lines to one atom such as oxygen, O, and nitrogen, N; this illustrates the *tetravalence* that Kekule proposed for carbon in all its compounds as a consequence of his first dream. In the formula for benzene, with six carbon atoms in a ring and one hydrogen atom attached to each, there are just

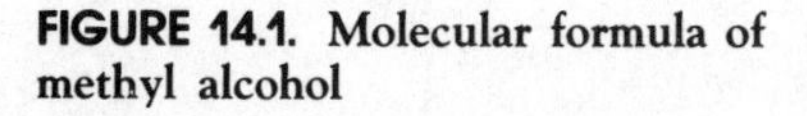
FIGURE 14.1. Molecular formula of methyl alcohol

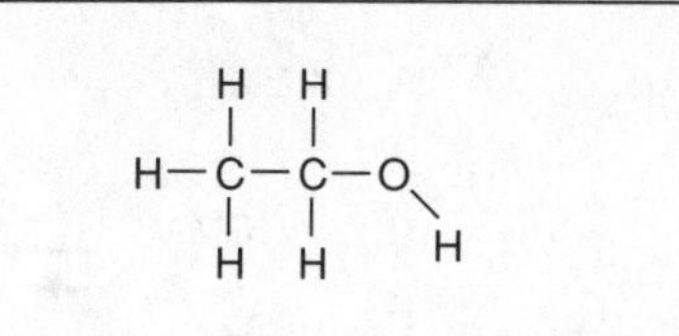
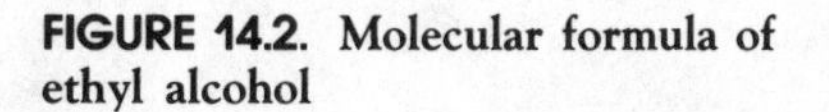
FIGURE 14.2. Molecular formula of ethyl alcohol

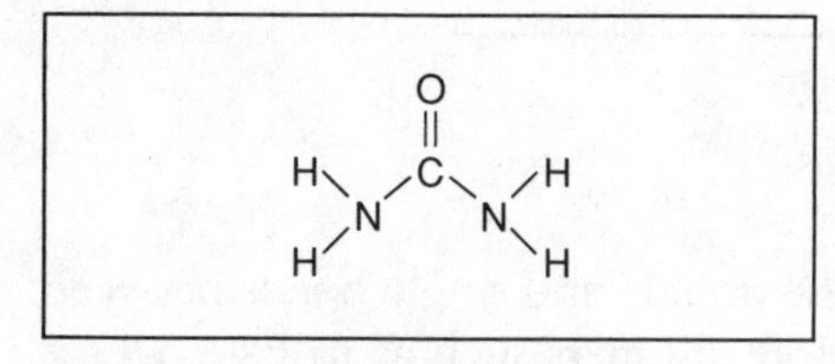
FIGURE 14.3. Molecular formula of urea

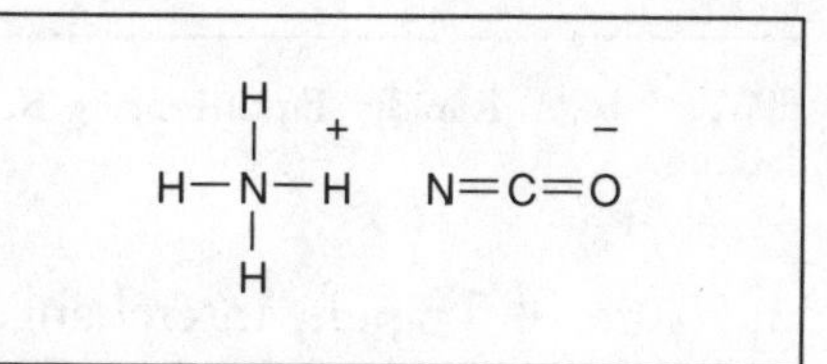
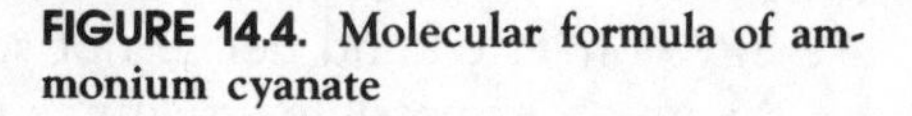
FIGURE 14.4. Molecular formula of ammonium cyanate

three lines (valence bonds) from each carbon atom to two other carbon atoms and to a hydrogen atom, unless some double bonds are inserted, as in Figure 14.5. So this was what Kekule did with "the unused valences" in his formula.

Although many chemists accepted and appreciated the ring structure, others pointed out a possible flaw in it. If two other atoms were substituted for two adjacent hydrogen atoms, there should be two different "isomers," one in which the two atoms (X in Figure 14.6) are attached to carbon atoms joined by a double bond (two lines, as in Figure 14.6a) and another in which the two atoms are attached to carbon atoms joined by a single bond (one line, as in Figure 14.6b).

Kekule modified his concept of the ring structure to account for the nonexistence of such isomers. He proposed that a ring structure such as

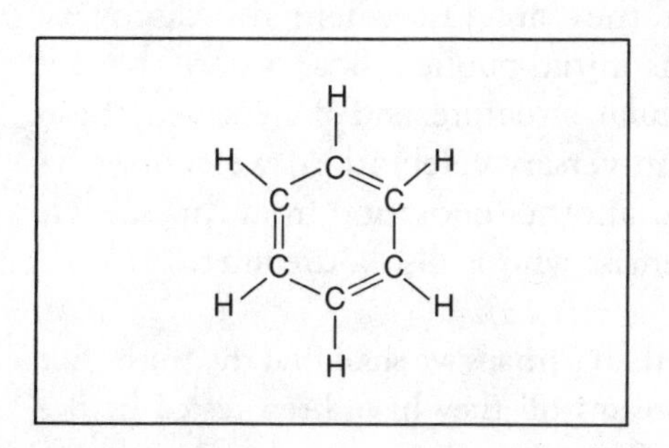

FIGURE 14.5. Molecular formula of benzene

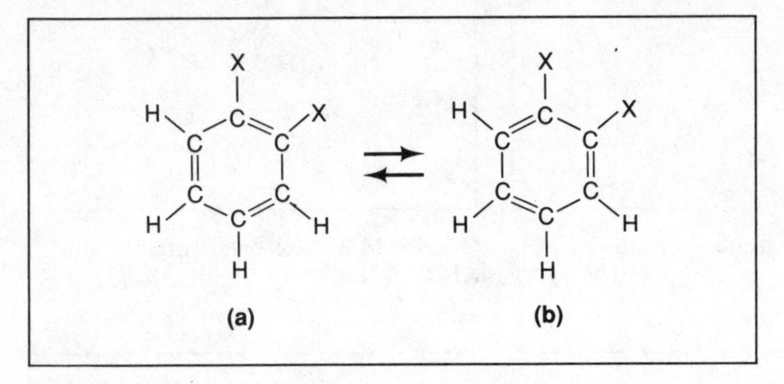

FIGURE 14.6. Rapidly Equilibrating Kekule Isomers

in Figure 14.5 rapidly interchanges the double and single bonds between the carbon atoms of the ring. Therefore, isomers such as in Figure 14.6 are interconverted and hence not separable.

Many other possible structures for benzene were proposed between 1865 and 1890, but none withstood the tests of experimental evidence as well as the Kekule structure. Kekule's view of the structure of benzene and of the thousands of related aromatic compounds resembles the modern view, which is based on the quantum mechanical concept of the electronic linking of atoms, although electrons (much less quantum mechanics) were not known, until years after the twenty-fifth anniversary celebration of Kekule's formula.

Some writers (*Chemical and Engineering News,* 1985, November 4 p. 22; January 20, 1986, p. 3) have criticized and even doubted Kekule's accounts of his dreams and of their role in his proposals of molecular architecture. In his publications in the 1860s Kekule did not refer to the dream origins. However, many scientists do not specify in their formal publications where they got their ideas, and they even present data in an order almost inverse to the actual sequence of events.

Accidents, imagination, and dreams have often been important ingredients in great discoveries, but they are just where the discovery *begins.* If Kekule was reluctant in his initial publications to state that he had dreamed his theories of molecular structure and if he saved these admissions for the speech at the anniversary celebration, we cannot be too surprised or suspicious. Consider another quotation from this speech that is characteristic of a great scientist who is also a dreamer:

> Let us learn to dream, gentlemen, then perhaps we shall find the truth. But let us beware of publishing our dreams till they have been tested by the waking understanding. (Benfey, p. 21)

Although Kekule could not have received a Nobel Prize because they were first awarded after he died, he was just the kind of person Nobel had in mind. A few months before his death Nobel said, "I would like to help dreamers, who find it difficult to get on in life."

The development of not only dyes, but also drugs such as sulfanilamide and aspirin, high-octane gasoline, synthetic detergents, plastics, and textile fabrics such as Dacron—all are outgrowths of the aromatic chemistry for which Kekule laid the foundation by his formula for benzene.

▼ POSTSCRIPT

In 1921 physiologist Otto Loewi discovered the humoral transmission of nerve impulses via chemical substances. According to U. Weiss and R. A. Brown (*Journal of Chemical Education,* September 1987 p. 770), the idea occurred to Loewi in a dream, not once but twice. After awaking from the first dream, he fell asleep again and could not remember it well when he awoke in the morning. Awaking from the second dream, Loewi went directly to his laboratory and performed the simple, but critical, experiments suggested in the dreams. Like Kekule, Loewi did not immediately publish the origin of the idea on which he acted. However, unlike Kekule, he did describe it at once to friends and family, and so the events were well documented. Loewi's daughter (who is the wife of Weiss, the first author of the 1987 article) remembers her father's description of his dreams and their consequences; she also remembers Loewi's co-workers predicting that his dream-initiated discovery would bring him a Nobel Prize. They were right. He shared the Nobel Prize in medicine or physiology with H. H. Dale in 1936, although in his Nobel lecture he did not mention the dreams.

Weiss and Brown go on to say that "the story of the background of Loewi's discovery proves quite conclusively that ideas for highly significant scientific research can indeed appear during sleep, as Kekule had related." They also mention that they found a statement by Hermann von Helmholtz, the great nineteenth century philosopher and physiologist, that fruitful ideas "often . . . came in the morning upon awakening."

In my experience, imagination and memory are most active in the dream state or semi-dream state (daydream?). I have seldom or never had a momentous idea while sitting in my office at the university. Such ideas are more likely to come in the early hours of the morning (as Helmholtz said), on an airplane or bus, on a pleasant or even a boring walk, in the shower, or while enjoying a musical concert.

Melvin Calvin (who won a Nobel Prize in chemistry in 1961 for explaining photosynthesis by plants) tells how the key to the puzzle came to him:

> I would like to describe the moment (and, curiously enough, it was a moment) when the recognition of one of the basic facets in the photosynthetic carbon dioxide cycle occurred. One day I was waiting in my car while my wife was on an errand. I had had for some months some basic information from the laboratory which was incompatible with everything which, up until then, I knew. I was waiting, sitting at the wheel, probably parked in the red zone, when the recognition of the missing compound occurred. It occurred just like that—quite suddenly—and suddenly, also in the matter of seconds the cyclic character of the path of carbon became apparent to me . . . in a matter of 30 seconds. So there is such a thing as inspiration, I suppose, but one has to be ready for it. (*Journal of Chemical Education*, September 1958, p. 428)

Charles H. Townes (Nobel laureate in physics, 1964), said "the laser was born one beautiful spring morning on a park bench in Washington, D.C. As I sat in Franklin Square, musing and admiring the azaleas, an idea came to me for a practical way to obtain a very pure form of electromagnetic waves from molecules." (*Science '84*, Vol. 5 p. 153.)

Neurobiologist Roger Sperry (California Institute of Technology professor and Nobel Laureate in medicine or physiology, 1981) studied epileptics whose brain hemispheres had been surgically separated. His research indicated that ideas and inspirations that come under circumstances like those described here come from the right side of the brain, and some persons have suggested that right-brain thinking can be consciously encouraged and even taught. (Englebardt, *Reader's Digest*, February 1988 p. 41)

Nevertheless, as valuable as dreams and daydreams may be, I agree with the last quotation from Kekule above—dreams must be reviewed and tested in the harsh light of day. An idea that comes as a flash in the night (or day) might require days, months, or even years of hard and brilliant work to come to practical fruition.

I have included these stories of discoveries through dreams as serendipitous because they have often been considered in this category. Whether or not dreams and inspirations are accidental might be debatable, but if a person acts upon them to make discoveries as valuable as those of Kekule, Loewi, Calvin, and Townes, they certainly must be counted as fortuitous.

15 ▼

NOBEL: The Man, the Discoveries, and the Prizes

▼ Many of the serendipitous discoveries described in this book have led to Nobel Prizes. The man whose fortune and legacy established the Prizes also was blessed by serendipity. The best known of his discoveries, dynamite, has been disputed as to whether it was accidental or carefully planned. In either case, the invention of the most powerful explosive known before the atom bomb led to Nobel's enormous wealth and subsequently, owing to Nobel's remarkable character, to the founding of the prizes.

In spite of the fame of the Nobel Prizes and their recipients as the news is flashed around the world each year, the general public knows little about Nobel, his life, inventions, and intentions. Nobel is worthy of being known as a pioneering inventor and industrial giant. In spite of acquiring great wealth and international renown before his death, Nobel's life was marked by ill health, tragedy, loneliness, and depression. Having developed explosives capable of greater destruction of life and property than could previously have been imagined, he sought by his

▼ *Alfred Nobel, 1833–1896*

legacy to champion for posterity the most precious cultural values of mankind in science, medicine, literature, and peace.

Alfred Bernhard Nobel was born in Stockholm in 1833, the same year that his father, Immanuel Nobel, went bankrupt. His health was fragile from birth; he survived only because of the special care his mother gave him. Of the four sons that lived past childhood, Alfred was clearly his mother's favorite, and he returned her love in special ways throughout his life. The father was a self-taught inventor and engineer, who put his own interests above those of family. When his business in Stockholm failed, he left his wife and three sons to go to Finland and then St. Petersburg, Russia, to escape debtors' prison and to try to recoup his financial losses.

After several years, Immanuel moved his family to St. Petersburg, where he was prospering in military explosive work for the Russian government and was even part owner of a workshop and foundry. Alfred had only two years of formal schooling in Stockholm, where he received, however, the highest marks in his class. In Russia, because of their father's financial success, Alfred and his older brothers were taught by tutors, but this situation was short-lived. With the end of the Crimean War in 1856, the Russian government's commitment to Immanuel Nobel's pioneering work on explosive mines for the navy ended and the elder Nobel experienced his second bankruptcy. This time he returned to Sweden with his wife and youngest son, Emil, leaving Alfred and his two older brothers, Robert and Ludvig, in Russia. Alfred had studied chemistry with a Russian professor, and he had gained much practical mechanical and engineering knowledge in his father's factory. He went to Paris to try to borrow money to save the factory in St. Petersburg, but he returned empty handed.

In 1861 Alfred was back in Sweden helping his father produce nitroglycerin, a new liquid explosive that had been first prepared by an Italian chemist, Ascanio Sobrero, a few years earlier. On a second trip to Paris he succeeded in obtaining a loan that allowed production to begin on a small scale in Sweden. Two years later, when Alfred was 30 years old, he made and patented the first of his major inventions, a mercury fulminate blasting cap to set off the nitroglycerin explosive. This invention was not accidental; records indicate that it was the outcome of over fifty exacting experiments in his father's ramshackle laboratory and factory at Heleneborg, on the outskirts of Stockholm.

Catastrophe struck in September of 1864. An explosion leveled the little factory, killing five persons, among them Emil Nobel, Alfred's younger brother. This tragedy apparently triggered a stroke that felled Immanuel Nobel, the father, and made him physically incapacitated for the remaining eight years of his life. It also gave Alfred a compulsion to discover a way to make nitroglycerin safe to manufacture, transport, and use. The whole responsibility for the family business now rested on him. He convinced the Swedish State Railways that nitroglycerin was much superior to black powder for blasting the mountain tunnels for the developing railway system.

Because of the explosion in the plant that killed his brother, no one wanted a nitroglycerin plant close by, and, in fact, Stockholm would not allow a plant within its limits. Not to be daunted, Alfred located his factory on a barge anchored in the lake beside the city. Soon after, with the backing of a wealthy Stockholm merchant, J. W. Smitt, a factory was built in an isolated area near Stockholm, and nitroglycerin was produced there for over fifty years. A second plant was built in Kruemmel, Germany, near Hamburg.

This was the beginning of a great expansion in the use of nitroglycerin, not only in Sweden, but throughout the world. For example, it was the key to the building of the Central Pacific railroad across the Sierra Nevada in the United States. Although accidents sometimes occurred, the tremendous saving of time and money in mining and tunneling operations led to acceptance of the enormously powerful liquid explosive for a time.

This period of success did not last long, however. Partly as a result of inadequate knowledge or disregard of instructions, but more often because of the inherent instability of nitroglycerin, reports of disastrous explosions began to come from around the world. In its sensitivity to explosion, nitroglycerin could well be described "capricious," sometimes exploding when tickled with a feather and at other times capable of being abused or used in extremely inappropriate ways with no difficulty. For example, there are reports of its being used to lubricate and polish leather boots and to grease the axles of carriages.

The final blow came with the destruction of the Kruemmel plant by an explosion in 1866. This led Nobel to pursue even more vigorously a means of stabilizing nitroglycerin. However, we find two different accounts of the successful solution to the problem that culminated in the invention of dynamite.

There is no disagreement that Nobel sought various means of taming nitroglycerin. He patented a method of adding to it methyl alcohol, which could be removed by washing with water when the explosive was to be used, but this proved to be impractical. He next tried to use solid powdered or fibrous material, such as sawdust, charcoal, paper, and even brick dust—all of which were unsatisfactory. The combustible materials were likely to be set on fire by standing with nitroglycerin, and the inert materials like brick dust reduced its explosive power.

According to one version of the discovery, a metal container of nitroglycerin was found to have sprung a leak and the liquid had soaked into the packing between the metal cans. This packing material was kieselguhr, a cheap, light, porous mineral that was widespread in northern Germany. Nobel happened to observe the pasty mixture produced by the leak from the damaged container. Presumably it occurred to him to test this material, and he found that it could be pressed into a compact solid that retained the explosive power of the liquid, but was safe and reliable until set off by a blasting cap. If this story were true, it would represent a classic case of pseudoserendipity. Nobel found by accident a solution that he had been seeking ardently.

There is, however, convincing evidence against this version of the birth of dynamite by accident. Nobel himself vociferously denied it, and maintained that he arrived at kieselguhr as the ideal adsorbent after careful scientific experimentation. There seems no reason to disbelieve Nobel's rejection of the accidental aspect of the discovery. Nobel was by all accounts rigidly honest, in spite of having to contend repeatedly with avaricious and unscrupulous business adversaries.

Nobel, however, made another discovery some years after inventing dynamite. He discovered blasting gelatin somewhat by design and somewhat by accident. Nobel himself described the events in 1875 leading to the discovery of blasting gelatin. Working in the laboratory one day, he cut his finger on a piece of glass. As was common practice at that time, he applied collodion to the injured finger. Collodion is a viscous solution of cellulose nitrate in ether and alcohol. Such solutions were used to form temporary coverings over wounds when the solvents evaporated. These coverings were highly flammable; cellulose that was more highly nitrated was actually a powerful explosive and was known as guncotton.

Unable to sleep that night because of pain in the cut finger, he began to ponder a problem he had considered earlier without coming to a satisfactory solution: how to combine nitrocellulose and nitroglycerin to produce an explosive more powerful than either but as safe as dynamite. He had experimented with guncotton and had so far been unable to combine it with nitroglycerin. With the collodion on his aching finger, it occurred to him that a lower degree of nitration such as that in the collodion might allow it to be better incorporated into a mixture with nitroglycerin.

He hurried down to his laboratory at 4:00 A.M. and began experimenting with different proportions of collodion and nitroglycerin. By the time his assistant arrived in the morning, Nobel was able to show him a clear jelly-like combination of the two most powerful explosives then known. Tests proved that the combination was indeed more powerful than either single component. After many carefully planned and executed experiments to determine the optimal formula for power and safety, Nobel patented blasting gelatin in 1875 in England and in 1876 in the United States.

The idea that led to Nobel's discovery of blasting gelatin was triggered by the accidental cut that brought collodion to Nobel's attention at a time when his mind was prepared to see a connection between collodion and the problem he was pondering. The other facet of serendipity illustrated here is the recognition of the possible significance of the idea and the immediate action taken by Nobel in following the lead suggested.

After the development of safe forms of nitroglycerin represented by dynamite and blasting gelatin, and several other modifications patented by Nobel, his explosives business expanded tremendously into both military and nonmilitary applications. The great tunnels blasted through the Swiss Alps—St. Gotthard, Simplon, Arlberg—could not have been made without the incredible force of the new gelatin dynamites.

▼ POSTSCRIPT

As Alfred Nobel became enormously wealthy, he remained, nevertheless, lonely and cynical. The one fact that most people know about him is that he left a tremendous fortune that became the source of prizes given annually in recognition of exceptional activities in various fields. Perhaps you have wondered why this fortune was directed to these ends rather than being passed to family descendents.

Poor health followed Nobel throughout his life; mental depression was also a frequent, if not constant, problem. He never married and, at the time of his death in 1896, little was known about his involvement with any woman other than his mother. This led to rumors that he was homosexual. Evidence to the contrary came from the remarkable disclosure more than 50 years later that there were actually three other women important in his life. Information, especially about one of them, was kept secret in the files of the Nobel Foundation until 1950, out of regard to persons who were by then deceased.

Apparently the first woman was a girl Nobel met in Paris when he was 18. In his early life Nobel wrote poems, and one of these speaks of a girl "good and beautiful" who returned his love and gave him, whose life until then "had been like a dreary desert," great happiness so that they became "a heaven to one another." Her sudden death brought him perhaps his first bitter disappointment. This early tragic love affair of the young and sensitive Nobel seems to have shaped his life.

When Nobel was 43 and in Paris again, he needed a secretary and housekeeper. His advertisement was answered by Countess Berta Kinsky, a wise and charming young woman of a noble but impoverished Austrian family. She came to Paris to take the position with Nobel because she was in love with a young aristocrat in Vienna, Arthur von Suttner, and his family strongly opposed their marriage. Berta was beautiful, intelligent, gifted in music and languages—in short, all the qualities that appealed to the shy and reticent Alfred Nobel. Although it seems certain that Nobel had no romantic aspirations in hiring a secretary, he almost immediately fell under her spell and is quoted as asking her "if her heart were free." She answered that it was not, and undoubtedly Alfred was disappointed. Their brief encounter led to a lifelong friendship, however.

After a very short time Berta returned to Austria and married von Suttner, whose family finally became reconciled to the marriage and accepted her. Berta von Suttner became a strong advocate of international peace. Nobel shared this concern, although he differed with her as to the means of accomplishing the ideal. He kept up a steady correspondence with Berta, and she undoubtedly had a strong influence on him, which was reflected in the setting up in his will of a Peace Prize. Berta von Suttner was the fifth recipient of this prize, in 1905.

Shortly after Berta left Paris, the third woman entered Nobel's life—probably his greatest love and his greatest disappointment. In the autumn of 1876 when Nobel was 43, he went to a flower shop in a health resort in Austria to buy a bouquet for the wife of a business acquaintance who was to be his hostess. There he met Sofie Hess, a petite 20-year-old beauty from a working-class Jewish family in Vienna. Their extensive correspondence and other documents disclosed in 1950 reveal an unusual love story of the intelligent, cultivated, disciplined, wealthy man who wanted companionship and relaxation in a comfortable home and the uneducated, undisciplined, charming young beauty who wanted only to enjoy life to the fullest.

Nobel set her up in Paris in a fine apartment with servants, and later in a villa at Ischl, near Vienna, but he left her alone much of the time while on business trips around the world. This arrangement was doomed, but 18 years passed with some periods of happiness before it ended as might be expected. Madame Sofie Nobel, as she was addressed in hundreds of letters from Alfred (although a marriage was never made official) turned to young admirers in fashionable resorts around Europe. Nobel tried to create an intelligent and cultured woman out of a pleasure-loving and spoiled minx, but without success.

Finally, Sofie told him that she was expecting a child by a young Hungarian officer. Nobel then gave up his efforts to reform Sofie, resolved never to see her again, but provided her with a liberal annuity. Sofie married but did not live with her Hungarian officer, and both he and she tried to extort money from Nobel until his death in 1896. Even then she was not satisfied; she threatened to sell the publishing rights of Nobel's 216 letters to her if she was not given more money than the will prescribed. After considerable negotiation, an agreement was reached by which Sofie turned over all of the letters in return for continuation of the annuity she had been receiving from Nobel during his lifetime.

Alfred Nobel died in his villa at San Remo, Italy, in 1896. He had several heart attacks in his later life and suffered frequently from angina pectoris. Shortly before his death he wrote in a letter, "it sounds like the irony of fate that I should be ordered to take nitroglycerin internally. They call it Trinitrin so as not to scare pharmacists and public." In his last two years his depression improved, probably for several reasons. He had been able to break off with Sofie and yet make her financially secure; he was happy with his home in San Remo, and with the purchase of the Bofors plant in Sweden, where he fitted up private quarters in an adjacent manorhouse and a laboratory for his experiments; and he had found a reliable new young assistant, Ragnar Sohlman (who later became the executor of his will).

In the autumn of 1895 Nobel spent two months in Paris, where he worked out the details of the will that became the basis of the Nobel Foundation and the prizes. He did this in his own hand in Swedish without help from lawyers. Because of this and the size of the estate, the will came in for quick criticism, and several years were required before his bequests could be implemented. At the time of his death, Nobel had no wife or children; his mother and all of his brothers were dead. Reasonable settlements were made to nephews and their families. Then the will stated that:

> The capital shall be invested by my executors in safe securities and shall constitute a fund, the interest on which shall be annually distributed in the form of prizes to those who, during the preceding year, shall have conferred the greatest benefit on mankind. The said interest shall be divided into five equal parts, which shall be apportioned as follows: one part to the person who shall have made the most important discovery or invention within the field of physics; one part to the person who shall have made the most important chemical discovery or improvement; one part to the person who shall have made the most important discovery within the domain of physiology or medicine; one part to the person who shall have produced in the field of literature the most outstanding work of an idealistic tendency; and one part to the person who shall have done the most or the best work to promote fraternity between nations, for the abolition or reduction of standing armies and for the holding and promotion of peace congresses.

One of Nobel's major aims, that of freedom from national bias, was criticized almost at once when the will was made public. The Swedish press contended that it was unpatriotic of a Swede to overlook Swedish interests and favor international activity, and Nobel's designation of the Norwegian Storting (Parliament) to award the Peace Prize was especially distasteful because of the strained relations between Sweden and Norway at the time. However, after stormy discussions between the executors and the family over several years, the Nobel Foundation was established in virtual accord with Alfred Nobel's wishes, with the Swedish government in control of the administration of the awards, but with no influence over the nomination or selection of prize winners. The first awards were made in 1901.

Alfred Bernhard Nobel hoped by the bequests from his huge fortune to accomplish what he had not been able to do in his lifetime: to encourage what would have "the greatest benefit" to mankind, especially peace and "fraternity between nations." Having invented and promoted the most powerful military explosive known, he hoped it would prevent war. In 1892 Berta von Suttner asked Nobel to join her at a Peace Congress in Switzerland. He declined and replied: "My factories may end war sooner than your Congresses. The day when two army corps will be able to destroy each other in one second, all civilised nations will recoil from war in horror and disband their armies."

There is a more modern parallel: The Federation of Atomic Scientists, formed after the development of the atomic bomb (which surpassed the power of Nobel's dynamite and blasting gelatin by a degree unimaginable by him) by some of those who participated in that project, has evolved into the Federation of American Scientists (also F.A.S.), which has as its main objective international arms control!

16

Celluloid and Rayon: Artificial Ivory and Silk

Celluloid

The first successful synthetic plastic was celluloid, which was developed to replace ivory in billiard balls. In 1863 there was a shortage of ivory, the material of choice for billiard balls, because of depletion of the herds of wild elephants in Africa. (Isn't it surprising to learn that what is a serious problem today was also of concern over a hundred years ago!) A major manufacturer of billiard balls offered a prize for an ivory substitute that could be used to make billiard balls.

A New Jersey printer named John Wesley Hyatt and his brother Isaiah began experimenting with various materials. One of these was a mixture of sawdust and paper bonded together with glue. When John Hyatt cut his finger while engaged in this work, he went to a cupboard to get some collodion to protect the wound. (Collodion, a form of cellulose nitrate dissolved in ether and alcohol, was popular for this purpose at that time. Alfred Nobel's similar experience, one that led him to the invention of blasting gelatin, is described in Chapter 15.) To his surprise, he found that the bottle of collodion had overturned, spilling the contents; the solvent had evaporated, leaving a hardened sheet of cellulose nitrate on the shelf. Hyatt realized that this material might make a better binder for his sawdust and paper mixture than the glue he was using.

After some experimentation, Hyatt and his brother found that cellulose nitrate and camphor, mixed with alcohol and heated under pressure, made a plastic apparently suitable for billiard balls. Nobel made blasting gelatin from cellulose nitrate combined with nitroglycerin. The camphor must have modified the explosive nature of the cellulose nitrate considerably; however, billiard balls made of celluloid did explode occasionally!

▼ *An advertisement for celluloid collars, around 1875*

Hyatt and his brother did not win the prize for an ersatz billiard ball, perhaps because the balls they made tended to explode. But they did patent their plastic made of cellulose nitrate and camphor in 1870 under the name "Celluloid," and it became popular for other applications. In the late nineteenth century, it was used for collars and cuffs of men's shirts. It was molded for dental plates, knife handles, dice, buttons, and fountain pens. More modern plastics have largely replaced it, but I remember as a boy identifying it as the material from which small pocket calendar cards were made—by its odor of camphor.

Rayon

Spilling a bottle of collodion also provided the idea for the first successful substitute for silk. When Pasteur was trying to save the French silk industry from a disastrous epidemic among silkworms, a young chemist named Hilaire de Chardonnet assisted him. From his experience with the silkworm problem, Chardonnet became convinced that an artificial substitute for silk was highly desirable. While working in his darkroom with photographic plates in 1878, he spilled a bottle of collodion. He did not immediately clean up the spill, and when he did, he found a tacky viscous liquid left by partial evaporation of the solvent, which produced long thin strands of fiber as he wiped it up. These fibers resembled silk enough to encourage Chardonnet to experiment further with the collodion.

Within six years after the accidental spill, Chardonnet produced an artificial silk. He derived the collodion from a pulp of mulberry leaves, (the natural food of silkworms!) dissolved in ether and alcohol, drew out filaments of the fiber, and coagulated them in heated air. Cloth made from the new synthetic fiber was displayed so successfully at the Paris Exposition in 1891 that financial backing was obtained immediately. The new fiber was called "artificial silk" until about 1924 when the name *rayon* was first used.

▼ POSTSCRIPT

Chardonnet's rayon was highly flammable. Other processes were later invented to convert cotton into silk-like fibers that were not flammable. Cellulose nitrate, the original rayon, is no longer used as a textile fiber. Cellulose nitrate was also once used for photographic film, both ordinary still film and commercial movie film. Its flammability was responsible for several disastrous fires in movie theaters; if a projector malfunctioned so that the film remained stationary in the path of the intense light for more than a few seconds, it would catch fire. The "safety film" that replaced it many years ago was cellulose acetate.

The fundamental difference in appearance of cotton and silk textiles lies in the filaments from which each is woven. The filaments of cotton are fuzzy, whereas the filaments of silk are smooth, just as they come from the silkworms. These smooth filaments give the silky sheen to silk. Chardonnet's rayon simulated silk because the cellulose (cotton or wood fiber) was converted into a chemically different form (cellulose nitrate) that could be dissolved in ether and alcohol, and smooth filaments could then be pulled or extruded from the viscous solution.

The most successful newer rayons are xanthate rayon and acetate rayon. Xanthate rayon takes its name from a process in which cellulose is converted into a chemically different and soluble form (cellulose xanthate). The viscous solution of cellulose xanthate is extruded through fine holes as smooth filaments into a chemical bath that converts the cellulose xanthate back into cellulose. The overall effect is to change the physical shape of fuzzy cellulose fibers into smooth silk-like fibers, and this rayon is actually regenerated cellulose.

Acetate rayon is more like the original Chardonnet rayon. Cellulose is converted into the acetate ester, which, like the nitrate ester, is soluble and can be extruded into smooth filaments. However, in contrast to cellulose nitrate, the cellulose acetate is not flammable. The textile industry now uses the generic term *acetate* for acetate rayon to prevent confusion with the other major type of rayon, xanthate rayon. If the label says simply "rayon," the material is probably xanthate rayon. It may be appropriate to mention a caution for anyone who may work in a laboratory where organic solvents like acetone may be used. "Acetate" is somewhat soluble in these solvents, and clothing made of it should not be worn in environments where contact with them is possible. Ordinary "rayon" (xanthate rayon) is chemically identical with cotton, and thus is impervious to organic solvents.

Various synthetic fibers have since been developed that are more like silk than rayon. They may even be superior to silk in some ways. The nylon that Wallace Carothers and his group at Du Pont developed in the 1930s is one of these fibers. New generations of nylon and other synthetic fibers, such as the polyesters, have been developed and have displaced rayon for many textile uses. (Chapter 25 describes the discovery of nylon and also mentions the polyesters Terylene and Dacron.) But Chardonnet's artificial silk started attempts to duplicate or improve on nature's silky textile.

▼▼▼▼▼▼▼▼▼▼▼▼▼▼▼▼▼▼▼▼▼▼▼▼▼▼▼▼▼▼▼▼

17 ▼

FRIEDEL AND CRAFTS —a Laboratory Accident Spawns New Industrial Chemistry

▼ The Friedel–Crafts reaction was named for the two chemists, Charles Friedel and James M. Crafts, who observed an unexpected result of an experiment in Friedel's laboratory in Paris in 1877. Friedel and Crafts recognized the potential practical importance of their accidental discovery; they secured patents in both France and England on procedures for preparing hydrocarbons and ketones. Their judgment was accurate. Probably no other organic reaction has been of more practical value. Major processes for the production of high-octane gasoline, synthetic rubber, plastics, and synthetic detergents are applications of "Friedel–Crafts chemistry."

James Mason Crafts was born in 1839 in Boston. After graduating from Harvard University at age 19, Crafts spent a year studying mining engineering, and, while studying metallurgy in Freiburg, Germany, became fascinated with chemistry. He worked in the laboratories of Robert Wilhelm Bunsen in Heidelberg and Charles Adolphe Wurtz in Paris. In the Wurtz laboratory he met Charles Friedel, and the two began collaborating in 1861.

In 1865 Crafts returned to the United States. After a brief stint as a mining inspector in Mexico and California, he became professor of chemistry at Cornell University, which had just started. Three years later, he moved to M.I.T., where he introduced numerous improvements in teaching and in the laboratories. In 1874 because of poor health he returned to Paris, where he and Friedel resumed their collaboration in the Wurtz laboratory.

When Crafts left M.I.T., he expected to return soon. Because of the change of climate or, perhaps, the excitement of the discovery he shared with Friedel, his health improved dramatically, but he remained in Paris for 17 years. Between 1877 and 1888 Friedel and Crafts produced over 50 publications and patents related to the reactions of aluminum chloride with organic compounds.

Upon the death of Wurtz in 1884, Friedel succeeded him as professor of organic chemistry and director of research at the Sorbonne. Friedel was a founder of the Chemical Society of France and served four terms as President of that Society.

In 1891 Crafts returned to M.I.T., where he was reinstated as a professor and in 1897 was elected president. He set out to make M.I.T. equivalent to European institutions by upgrading the teaching and research standards. After three years, he resigned his administrative position to spend more time on experimental research, in which he continued until his death at age 78 in 1917.

Let us return to the accidental research event that did more to engrave the names of Friedel and Crafts in scientific and industrial annals than the high academic and administrative positions they both held. They were attempting to prepare amyl iodide by treating amyl chloride with aluminum and iodine. The reaction produced large amounts of hydrogen chloride and, unexpectedly, hydrocarbons. They then found that aluminum chloride in place of aluminum gave the same unexpected results. Earlier workers had reported somewhat similar results from reactions of organic chlorides with certain metals (zinc, for example), but had not explained them or implicated the metal chloride as a reactant or catalyst. Friedel and Crafts first showed that the presence of the metal chloride was essential.

In describing their discovery to the Chemical Society of France, they reported, "With a mixture of chloride and hydrocarbon, the formation is established, in good yield, of hydrocarbons from the residues of the hydrocarbon less H and from the chloride less Cl. It is thus that ethylbenzene, amylbenzene, benzophenone, etc., are obtained."

Friedel and Crafts saw that their unexpected result promised the possibility of synthesizing a wide variety of hydrocarbons and *ketones* (benzophenone is an example of a ketone, the second important class of compounds that could be made by their new reaction), and they proved this by experiment. In the ensuing years, Friedel and Crafts' voluminous research papers and patents established a new area of research and practice in organic chemistry and laid the foundation for some of the most important modern industrial chemical processes.

Friedel–Crafts chemistry might seem technical and complex, but it has touched our lives in many important ways. Winston Churchill, referring to the winning of the war over Britain by fighter pilots, said "Never in the field of human conflict was so much owed by so many to so few." The victory in the air war was due not only to the skill and daring of the British pilots, but to the superiority of their aviation gasoline. The German fighter planes were superior to the British and American planes, but their fuel was not. The aviation fuel used in the British and American planes gave them a critical performance edge. This fuel was a direct outgrowth of Friedel–Crafts chemistry. It contained toluene and other alkylated aromatic hydrocarbons.

Similarly, synthetic rubber was vital to the ground war effort in World War II after the Allies were cut off from sources of natural rubber. Tires for cars, trucks, and planes were essential. Because of cooperation between government officials and industrial and academic scientists and technicians, synthetic rubber was produced very quickly. It was made from styrene. This synthetic rubber was called GRS, which stood for government rubber, styrene type. It was a copolymer of styrene and butadiene (C_4H_6). A copolymer is a macromolecule produced by polymerizing a mixture of two monomers; styrene and butadiene are the monomers, in this case.

A polymer made from acrylonitrile (C_3H_3N), butadiene, and styrene (ABS) has had wide application. It has the toughness of a styrene/acrylonitrile copolymer and some of the flexibility of a styrene/butadiene copolymer. ABS polymers have been used for hand luggage and automobile parts, for example.

Styrene can be polymerized by itself. This polymer, called simply polystyrene, is very versatile. It can be molded into cases for radios, batteries, toys, and all kinds of containers. Because of its insulation properties and lightness, polystyrene foam ("styrofoam") is used for insulation in buildings and is molded into ice chests and disposable cups for hot and cold drinks. A recent development is the use of styrofoam in sculptures for outdoor advertising. Over four billion pounds of polystyrene products were produced in 1982.

Synthetic detergents have revolutionized the way we live today. We wash dishes and clothes with them; in contrast to soap, they work well in hard water. They are a major ingredient in shampoos. A typical example of a biodegradable synthetic detergent is sodium dodecylbenzenesulfonate. The 12-carbon sidechain is attached to a benzene molecule by a Friedel–Crafts alkylation reaction.

The phenol that is used in making aspirin (see Chapter 29) is produced from isopropylbenzene (also known as cumene), which is made by a Friedel–Crafts reaction of benzene with propylene. Each year over 40 million pounds of aspirin are produced in the United States, corresponding to 300 tablets for every man, woman, and child.

These are some examples of the practical applications that have been made in the century since the laboratory accident Charles Friedel and James M. Crafts observed in 1877 and interpreted through their sagacity to become a major discovery. In *Friedel–Crafts and Related Reactions* (1963), G. A. Olah and R. E. A. Dear wrote about the discovery by Friedel and Crafts: "It cannot be denied that many important scientific discoveries were made by chance, but it is essential that a person engaged in the research must have a keen sense of observation and creative ability if any discovery is to be developed and not remain unknown or passed over."

▼ POSTSCRIPT

A few years after I joined the faculty at the University of Texas at Austin, I spent a summer at the Oak Ridge National Laboratory learning how to use radioactive carbon (^{14}C) as a research tool in organic chemistry. Radioactive carbon is a by-product of the atomic energy program. Its atoms can be introduced into organic molecules and special instruments can detect their presence and even their positions because of their radioactivity.

When I returned to the University in Austin, I used this new technique in my research. The problem I chose for a ^{14}C tracer experiment was a Friedel–Crafts reaction that German chemists reported in 1892. Based on modern interpretations of Friedel–Crafts reactions, their results appeared to be questionable, and the experimental evidence was less than convincing. I saw a chance to confirm or deny their results using modern experimental techniques. Two such techniques were infrared analysis and gas chromatographic analysis; either of these would have given a dependable answer to the problem, and both were inherently simpler than ^{14}C radiochemical analysis. However, because of my interest in the new (to me) radiochemical ^{14}C technique, I decided to use this test of the old work. Stanley Brandenberger, a graduate student, agreed to carry out the experiments.

If we had used infrared or gas chromatographic analysis of the experiment, we would have found that the original work was correct, as far as it went. However, using the ^{14}C technique, we discovered an unrecognized molecular rearrangement. Exciting though this finding was, it had no practical significance, because the molecules labeled with ^{14}C we used were not in ordinary aromatic hydrocarbons. However, alerted by our discovery with the ^{14}C-labeled molecules, we looked for similar molecular rearrangements in ordinary aromatic hydrocarbon molecules, and we found them—including some of practical importance.

We were able to correct some incorrect reports in the chemical literature and to explain other puzzling results. Study of the "alkybenzene rearrangement" that we had inadvertently discovered occupied us for many years, and incidentally aroused my continuing interest in other examples of scientific serendipity.

18 ▼

How to Succeed in Archaeology Without Really Trying

▼ Mary Leakey, an archaeologist, the wife of an archaeologist, and the mother of an archaeologist, once said: "In archaeology you almost never find what you set out to find." Indeed, many famous archaeological discoveries have been made by persons who had no intention of finding anything of historical interest—that is, by serendipity. These discoveries have provided knowledge of ancient civilizations such as the Roman Empire (at Herculaneum and Pompeii) and the first Chinese empire (near Xi'an in central China); of prehistoric cultures as represented by the drawings in the caves at Lascaux and Aurignac in France and the ships of the Bronze Age found at the bottom of the Mediterranean Sea; of ancient and prehistoric humankind such as Tollond man of Denmark, Neanderthal man of Germany, and the Taung child of South Africa; and of religious history and documentation as brought to light by the Dead Sea Scrolls.

Digs That Produced Unexpected Results

Herculaneum and Pompeii. In 79 A.D., the volcano Vesuvius erupted, burying the cities of Herculaneum and Pompeii. Lava and ashes poured down so quickly and heavily that they instantaneously entombed the people and buildings of these neighboring cities.

In 1709, while digging a well in the farmland that had formed above Herculaneum, a peasant brought up fragments of sculptured marble. An Italian prince learned of this, bought the land, and brought workers to enlarge the vertical shaft and then dig horizontally. They found several intact female sculptures; the shaft had penetrated into Herculaneum's theater. The news of the buried cities spread and the Italian King Charles III hired a Spanish engineer to excavate and move every treasure that was portable to his private museum. During the Napoleonic wars of the early nineteenth century, the careless excavations by the French government were as bad as those by the Italians before them: frescoes and statues were removed from the temples and the exposed buildings were left to decay.

When King Victor Emmanuel II ascended the throne of Italy in 1860, fired by his view of the great Roman past, he encouraged careful excavations at Pompeii under the direction of Giuseppe Fiorelli, a professor who knew the history of Pompeii. Fiorelli developed an ingenious procedure for producing plaster casts of the human victims whose bodies had been encased in lava but had decayed during the centuries so as to leave hollow shells. After filling the shells with plaster and letting the plaster harden, workers carefully chipped away the pumice and ash "molds," leaving lifelike forms of the victims in the positions in which they had encountered death.

At present three-fourths of Pompeii's area has been uncovered and cleared, including two theaters and the civic forum. The rest of the city still lies buried beneath the houses and gardens of the present Italian town. The nearby brooding volcano still smokes from time to time. The last major eruption occurred in 1944.

China's Qin Tomb. In 1974, well diggers in the People's Republic of China discovered another ancient culture. Peasants dug up parts of life-size terracotta figures near the mausoleum of Qin Shihuangdi, who in 221 B.C. proclaimed himself the first emperor of a unified China. By his order the Great Wall was built to defend the northern boundaries of his empire from the Mongols.

Further excavation at the well site uncovered a series of underground vaults that proved to be one of archaeology's most stunning finds. The first pit, an area of almost four acres, contained 6,000 life-size terracotta soldiers and six four-horse chariots in 11 rows. Two years later a second vault or pit was found, with an area of 2.5 acres, and containing 1,400 figures of horses and men. At almost the same time, a third smaller pit was uncovered; it contained 73 soldiers guarding commanders who rode in chariots. These pottery warriors were not produced by an assembly line; each has a different face, suggesting that they were modeled after actual persons. The different physical features of the soldiers, typical of the minority nationalities from different regions of the empire, reflect the large number of conscripts that made up the army of the First Emperor. The uniforms and weapons have also provided much information about warfare in those times.

Over the largest vault has been built a covered museum. There visitors look down over railings to see the warriors of 2,200 years ago lined up in martial array on the vault floor 20 feet below ground level. Many of the soldiers and horses were damaged when the roof caved in sometime since the pottery army was first deployed, but repair and restoration continues today.

Taung Child. In 1924, workers were excavating for lime in the Taung cave near Johannesburg, South Africa, when one of them saw among the limestone rocks something resembling a cast of a small brain. He took it to the mine office, where a message was sent to Professor Raymond Dart, head of the Anatomy Department at the University of Witwatersrand. Soon Dart announced in the British journal *Nature* the discovery of an almost complete skull of a child, which he had freed from the limestone matrix found in the Taung cave. This child represented the oldest human ancestor known at that time—more than a million years old. (A spectacular hologram of the Taung child's skull is reproduced on the cover of the November 1985 *National Geographic.*)

Some might have thought that the Taung skull fell into the wrong hands when it was given to Dart. After all, he was young (31), rather inexperienced, and slightly inclined to scientific heresies. As it turned out, he was exactly the right person to turn an accidental find into a major scientific discovery. He had the perceptiveness, almost prescience, to recognize this ancient adolescent child as what some have called "the missing link" between our nonhuman and human ancestors. From evidence of the angle of carriage of the head, as deduced from the shape of the skull at the base of the cranium, Dart claimed that the Taung child walked upright. These and other ideas of Dart stirred up a tumult of controversy—even over the fact this ancient fossil was found in Africa, which was not thought to be the site of the oldest human ancestors. But subsequent work in Africa by a remarkable group of archaeologists (the Leakeys, Donald Johanson, and others) corroborated most of Dart's claims.

Neanderthal Man. In 1857, quarrymen digging in a cave near Neanderthal, Germany discovered a controversial "man." The "Neanderthal man" was actually an assortment of brown bones, including a skull. Soon thereafter Darwin published *The Origin of Species* (1859) and Thomas Huxley claimed that evolution by natural selection provided a theoretical framework for early human history. Some scientists dismissed the Neanderthal skull as that of an idiot, but Huxley pointed out that it had ape-like characteristics and provided a link between humans and apes. The view that the Neanderthal skull represented an early human was substantiated by discoveries in 1886 in a cave in Spy, Belgium, of similar skeletons, accompanied by chipped stone implements and animal bones of extinct subarctic species. Neanderthal man is now accepted as being over 100,000 years old.

Tollond Man. In 1950, two men digging for peat in Tollond swamp in Denmark uncovered a well-preserved corpse of a man, wearing only a leather cap and belt and with a serene expression on his face despite a leather rope twisted tightly around his neck. Archaeologists concluded that this man, who came to be known as Tollond Man, had been executed, probably as a religious sacrifice. Archaeologists, anatomists, and botanists analyzed everything from his cap to his last meal. His body had been so well preserved in the peat that the hair on his head was in good shape and a stubble of beard could still be seen. (Peat bogs contain tannic acid, which serves as a preservative, and the peat covers objects in such a way as to exclude oxygen, which would otherwise cause deterioration.) He had been about 30 years old when he died some 2,200 years ago.

Lindow Man. In August 1984, a commercial peat cutter working near the airport in Manchester, England, was about to throw a load of peat into a shredding machine. Some of the attached moss fell away from the peat, revealing a human foot. Archaeologists who were called in recovered one of the most perfectly preserved ancient human bodies ever found.

Studies revealed that this man, who came to be called Lindow Man, was a Celt and belonged to the ruling caste known as Druids. He, like Tollond Man, was apparently sacrificed in a religious ceremony as a result of losing a macabre lottery 2,200 years age. He had a serene expression on his face, indicating that he went willingly to his death, a brutal execution that included drowning, cutting his throat, crushing his windpipe with a leather thong, and bludgeoning his head.

Chemical analysis of the contents of his stomach and intestines disclosed a partly digested, badly scorched piece of cake. Dr. Anne Ross, an archaeologist specializing in Celtic history, concluded that this scorched cake was the losing "lottery ticket" that Lindow Man drew to seal his fate as the sacrifice to the Druid gods. Ross suggested that the burned cake was one section of a ground barley cake used in Druid ceremonies in which a portion of the thin, flat cake was allowed to scorch. Priests of the order would then break the cake into pieces, place them in a leather bag, and pass the bag around the circle. Each participant would take a piece; the unlucky man who drew the scorched piece would be sacrificed.

Archaeologists have suggested that the finding of this Celt in England has significance regarding the ancient history of Europe. They believe that the Celts may have dominated more of Europe than we thought, and that Celts rather than Germans ruled Scandinavia in the second and third centuries B.C.

Mexico City Aztec Disk. In 1978, workers for a utility company in Mexico City were digging a trench for an electric cable when they encountered a large stone disk. They recognized the carvings on the disk as images of dismembered parts of an Aztec moon goddess, who according to legend was slain by her brother, the war god of the Aztecs. The electric cable was forgotten as archaeologists moved in to excavate. For four centuries the Aztec city had lain buried beneath Mexico City. Now the ruins of the temple of Huitzilopochtli were found. It contained paraphernalia of over a hundred ceremonial offerings, including sacrificial knives, gifts of tribute, and, in one stone-lined pit, the skulls of 34 children from three months to eight years of age who had been sacrificed to Tlaloc, the rain god.

Mammoths in the Black Hills. In 1974, during excavation for a new housing development in the outskirts of Hot Springs, a community near the Black Hills of South Dakota, George Hanson, the operator of an

▼ *Mastodon tusks found during an excavation at a construction site in Austin, Texas*

earthmover, uncovered a cache of bones about 20 feet beneath the surface. Hanson's son reported the find to his former professor, Larry Agenbroad, who recognized the bones as those of Columbian mammoths. They had been trapped in a sinkhole 26,000 years ago. A building has now been built over the site so that visitors can see the fossilized bones in the positions in which the huge prehistoric animals died. Excavation still continues at the site one month each year.

Mastodons in Austin. In January 1985, while excavating for the foundation of a 22-story office building in Austin, Texas, a backhoe operator unearthed a well-preserved ivory tusk. Alton Briggs, an archaeologist employed by the contractor to be at the site during excavation (not uncommon practice, because archaeological artifacts are often found during excavations) identified the tusk as that of a mastodon, a prehistoric animal like the mammoth, related to the elephant. Mastodon fossils are rarer than those of mammoths; only a few hundred mastodons have been found, whereas over 3,000 mammoths have been unearthed. Tusks of three mastodons were found, as well as a rib, part of an upper left jawbone with two large teeth intact, spinal bones, a leg bone, and a foot bone. One of the two teeth, a molar, was so well-preserved that it was shiny after washing.

A University of Texas paleontologist, Ernest Lundelius, who worked with Briggs after the bones were discovered, said that this find was unusual: "It's not very common to find this much of the animal in a situation like this where you can pin down where it came from. Usually mastodon bones are dredged up from creek beds, where they might have ended up after being washed downstream." This site was thought to be the bed of a pond along the old flood plain of the Colorado River, which flows through Austin. Radio-carbon dating of the strata above and below where the mastodon bones were found set their age at about 15,000 years.

A Queen's Statue in Akhmim. In 1982, while on a lecture tour, I saw the beautiful statue of a favorite wife of the Pharoah Ramses II (reigned 1304–1237 B.C.) in Akhmim, a city on the Nile in upper Egypt. Workers had just discovered it while excavating for the foundation of a new building. The statue had lain buried in the ruins of an unknown temple for 32 centuries, yet the red color on her lips and the blue of her eye shadow were still visible on the smooth marble!

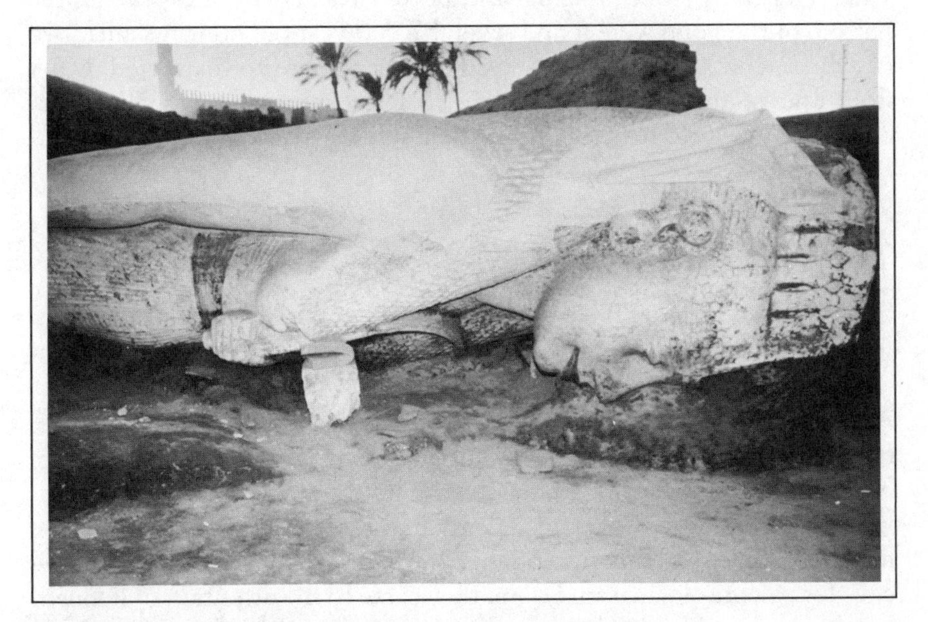

▼ *A statue of a Pharaoh's queen found during an excavation in Akhmim, Egypt*

Lakes That Dried Up

Biskupin Village. In 1933, a schoolmaster on an outing with his class in Poland noticed a number of evenly spaced wooden posts sticking out of a lake, and became curious about them. His investigations brought to light the village of Biskupin, which had been built on an island in the middle of the lake over 2,300 years earlier. The residents of the village had abandoned it owing to poor harvests and military invasions, and the site gradually disappeared under rising waters of the lake. When dredging of the rivers in the area lowered the surface of the lake in the 1930s, the island and its ancient village reappeared.

Excavations that began in 1934 displayed a carefully planned village covering about six acres, surrounded by walls and an embankment of logs. The well-preserved remains yielded weapons, tools, and pottery; archaeologists were able to reconstruct much of the village, including a row of houses joined by a common roof. The keen observation of the schoolmaster of the remains of the village emerging from the lake led to a valuable center for studies of a little known culture.

Titusville Skulls. In 1985, the intentional draining of a pond in Titusville, Florida, produced some unexpected results. In the peat at the bottom of the pond were found several 7,000-year-old human skulls, two of which contained well-preserved brains. Scientists extracted DNA from one of the brains; they hope to learn more about evolutionary changes in genetic structure from their studies.

Things That Just Turned Up

Some ancient artifacts have turned up on the surface of the earth more or less accidentally, where alert observers have noticed them.

Aurignac Cave. In 1852, a road worker pulled a human bone out of a rabbit hole near the village of Aurignac in the Pyrenees foothills of France. The curious worker enlarged the hole and found a cave containing the skeletons of 17 people, perforated shell disks, and the teeth of mammals. The road superintendant collected some of the contents of the cave for paleontological study, and they came to the attention of Edouard Lartet, a lawyer whose hobby was collecting fossil bones. A few years later, Lartet visited Aurignac; digging further around the cave he found tools made from flint and antler as well as the bones of extinct mammals.

These findings and the absence of metal artifacts convinced him that the Aurignac people had lived during the early Stone Age.

Ban Chang Artifacts. In 1966, while doing sociological work in Thailand, Stephen Young was walking down a road cut through a mound in the village of Ban Chang when he tripped on a root and landed face-to-face with a pot protruding from the earth. On the sides of the road cut he saw many more pots. He guessed that they were very old because they were unglazed. Although local villagers pillaged the immediate area severely, when Chester Gorman, an archaeologist from the University of Pennsylvania, arrived a few years later and dug beneath the middle of a nearby village street, he removed 18 tons of material for study. He arrived at a new theory about the prehistory of this part of Southeast Asia, because he could date the bronze jewelry and spear points as early as 2,000 B.C. According to the old theory, metallurgy first developed in the Near East and then spread to Southeast Asia as late as 500 B.C.

Folsom Indian Relics. In the fall of 1908, a cowboy named George McJunkin was riding along a gully near the town of Folsom, New Mexico, when he saw a bone protruding from the wall of the gully. When he pulled on it, a large bone came loose, and he realized it did not look like any of the cattle or buffalo bones he was familiar with. Many years later Jesse Figgins of the Colorado Museum of Natural History visited the Folsom site and decided that the bones might be those of a species of bison believed extinct since the end of the Ice Age. What was more exciting, along with these bones he found metal spearheads, some obviously embedded in the skeletons of the animals. Frank Roberts, Jr., from the Smithsonian, examined the artifacts and endorsed Figgins' theory that humans had lived in this part of America as early as the end of the Ice Age, about 10,000 years ago. This was about 5,000 years earlier than was thought before the discovery of the bones of the extinct bison.

Rosetta Stone. Perhaps the ultimate archaeological discovery coming from an artifact that turned up unexpectedly is the Rosetta Stone, the key that unlocked the history of ancient Egypt. A French soldier in Napoleon's army found the stone during repairs to Fort St. Julien near the town of Rosetta, on the bank of one of the western branches of the Nile in the delta region a few miles from the sea and about 30 miles west of Alexandria. Carol Andrews of the British Museum, where the stone now resides, describes the discovery:

▼ ***The discovery of the Rosetta Stone by Napoleon's soldiers***

The circumstances of its discovery in mid-July 1799 are in some doubt. According to one version, it was just lying on the ground, but the likelier account records that it was built into a very old wall which a company of French soldiers had been ordered to demolish in order to clear the way for the foundations of an extension to the fort later known as Fort Julien. . . .

The officer in charge of the demolition squad, a Lieutenant of Engineers called Pierre Francois Xavier Bouchard, and his officer companions are credited with having realized almost at once the importance of the Stone's three distinct inscriptions, namely that they were each versions of a single text in three different scripts. Since the last of the inscriptions was in Greek and could therefore be read, it was clear that it might be possible to use its translation as the key to the decipherment of the hieroglyphs in the first section. . . . On its arrival in Cairo in mid-August the Stone at once became an object of the deepest interest to the body of learned men whom Napoleon had taken with him to Egypt. . . .

On 24 August A.D. 394 on the island of Philae at the southern border of Egypt hieroglyphs were used apparently for the last time to inscribe the ancient Egyptian language. The final inscription on stone in demotic, the latest and most cursive of the three scripts employed by the ancient Egyptians, is dated less than 60 years later, to A.D. 452. . . . For all useful purposes during the next 1,370 years ancient Egypt was silent, for the art of reading her ancient scripts had been lost. There was no one to give voice to

> the countless hieroglyphic inscriptions which swarmed all over her monuments or to the texts in cursive hieratic and demotic crowded on papyri and flakes of stone and pottery. (Andrews, The British Museum Book of the Rosetta Stone)

There were many who contributed to the decipherment of hieroglyphs, but it was Jean-Francois Champollion, working from the Rosetta Stone, who established the foundations on which present knowledge of the language of the ancient Egyptians is based. With the hieroglyphs on the Rosetta Stone deciphered, the information from the inscriptions that "swarmed all over" the monuments gave a marvelous and complete picture of the civilization along the banks of the Nile through many centuries.

Nature Sometimes Lends a Hand

One variation of Murphy's Law is "Nature is by nature perverse." But not always. Nature has sometimes helped reveal past history to archaeologists. An excellent example, which must be considered pseudoserendipity is the story of Lucy because she was found by those seeking someone like her, but with an assist from Mother Nature. (The finding of the skull of the Taung child in South Africa was truly serendipitous, because the lime diggers had no intention of looking for fossils.)

Lucy. In November of 1974, anthropologist Donald Johanson and a graduate student, Tom Gray, were searching for ancient human fossils in north-central Ethiopia, but finding only the bones of animals. When walking along a gully they noticed a bone protruding from the eroded slope above them. Although it had been buried under layers of sediment and volcanic ash for millenia, a recent flash flood slicing down the gully had laid bare this bone. And not just one bone (the first was an arm bone), but many more; after three weeks of feverish excavation, Johanson and his colleagues found several hundred pieces of bones. These pieces proved to belong to a single individual—an adult female only three feet eight inches tall—who came to be known as "Lucy," a name taken from a Beatles song popular at that time.

Lucy stirred up great excitement because she represented the most complete and oldest prehistoric human ancestor known at that time. Radioisotopic dating placed her age at about three million years before the present, and the nature of her pelvic bones established that Lucy walked in an erect bipedal manner one or two million years before the Taung child lived.

Ozette Indian Relics. The village of Ozette is another example of how nature has helped in a pseudoserendipitous way to disclose long-buried remains. Brian Fagan described this finding in *The Adventure of Archaeology (1985).*

> Richard Daugherty, an anthropologist at Washington State University, had few written records to consult when he excavated buried remains at Ozette, a tiny abandoned coastal village in the northwestern corner of Washington. But in piecing together the history of the Makah Indians, who inhabited Ozette until half a century ago, he had their modern descendants to help him. . . . Tribal elders kept the legends of their ancestral home alive. One of the tales told of a great disaster—a huge mountain of mud that had buried their village long ago.
>
> Daugherty suspected that there was truth to the legend, and in 1970, when violent winter storms sent tides raging up the broad beach at Ozette, his suspicions were confirmed. Part of a bank washed away to reveal an abundance of artifacts—a canoe paddle, fishhooks of wood and bone, parts of inlaid boxes, a harpoon shaft, a woven hat—all dating to about the time of Columbus's arrival in America. They had survived the centuries under a layer of dense mud. . . .
>
> The Makah Tribal Council built a museum to house the cultural remains. . . . Say the Makah: "We. . . look in a special way at what [has come] from the mud at Ozette, for this is our heritage."

Even Bombs Have Helped

At the end of World War II, large areas of some of Europe's greatest cities—among these, London, Berlin, Rotterdam, Hamburg—lay in ruins from aerial bombs. Sad as it was, the devastation provided a unique opportunity for archaeologists. Among the rubble in London, the newly-formed Roman and Medieval London Excavation Council sought signs of Londinium, the riverside town founded by the Romans about 43 A.D.

A temple dedicated to Mithras, chief diety of a cult popular with Roman soldiers, was found. Just before the pile drivers of the construction workers were to destroy the temple in preparation for the foundations of an office building, public opinion prevailed upon the government to require the contractor to move the Roman shrine to a nearby location, where it continues to attract visitors interested in the city's past.

Boys and Caves

The curiosity of boys about caves they have found has on at least two occasions brought forth valuable archaeological information.

Lascaux Cave. In 1940, four boys were exploring the woods near the town of Montignac in southwestern France when they discovered a small hole in the ground. Curious, they enlarged the hole enough to crawl inside, and they found a narrow passage that led into a large underground cave. With the light of their oil lantern they were amazed to see brilliantly colored paintings of animals on the white limestone walls and ceiling of the cave. They reported their discovery to their former schoolmaster, who made a telephone call to the Abbe Henri Breuil, an expert on prehistoric art. Breuil came to see the cave paintings and proclaimed them to be authentically ancient. When news of the discovery reached the public, archaeologists, journalists, and sightseers came to the cave; they were admitted in small, carefully controlled groups. After the war the French government and the landowner provided a safer entrance and better security for the paintings. Since that time, thousands of tourists have come to Lascaux Cave. They now can view an exact replica that has been built adjacent to the original cave without danger of damaging the irreplaceable original artwork.

Dead Sea Scrolls. In 1947, a Bedouin boy was searching for a goat that had strayed among the barren cliffs bordering the northwest coast of the Dead Sea. He noticed a small opening in one of the cliffs and tossed a stone into it, only to hear the shattering of pottery inside. He was frightened away by the sound, but he returned with a friend and they managed to crawl inside the cave, where they found several large earthen jars. In some of these jars they found aged scrolls of parchment wrapped in linen. The boys took the scrolls back to their camp. Some of them were subsequently sold to the Syrian archbishop in Jerusalem, 13 miles to the west.

Although a series of experts who examined the scroll manuscripts pronounced them to be worthless, the archbishop sent them to the

▼ *Discovery of the Dead Sea Scrolls*

American School of Oriental Research in Jerusalem. It was there that the archaic forms of the Hebrew letters on the scrolls convinced Drs. John C. Trever and William Brownlee of the sensational nature of the documents. They made photographs of sections of one scroll (a copy of the Biblical book of Isaiah) and sent them to Dr. William F. Albright, an authority on Hebrew paleography at Johns Hopkins University. Dr. Albright promptly dated the manuscript at about 100 B.C. and called the scrolls "an absolutely incredible find." The paleographic dating was subsequently confirmed by carbon-14 dating. These scrolls antedated all but a few fragmentary Biblical manuscripts by 1,000 years.

Armed hostility between Arabs and Jews delayed further investigation by archaeologists until 1949, and then the initial searches for more scrolls and for information about the people who had prepared and secreted them failed. However, the impoverished Bedouin of the region saw the possibility of a new source of income (they had sold the original scrolls to the eager Jerusalem scholars). They began busily to scour the thousands of cracks and fissures in the barren desert wilderness of Judea

▼ *Cave No. 4 at Qumran near the Dead Sea*

near the Dead Sea. In 1952 they struck pay dirt in the vicinity of the ruin of Qumran, less than a mile from the original finding in Cave No. 1 by the Bedouin boy. Further intense activity by both Bedouin and archaeologists disclosed the remains of an ancient monastic settlement of an ascetic Jewish sect known as Essenes. It was they who produced the scrolls and hid them, apparently intending to return for them after the Roman persecution.

Thousands of fragments of nearly 400 separate scrolls have been found, including portions of every book of the Old Testament except Esther. Decades may pass before the full import of the scrolls can be assessed, but already they have filled important gaps in our knowledge of the Bible. They give us a new understanding of the religious climate into which Jesus was born, and for the first time, the mysterious Essenes stand revealed to us.

Sponge Divers

Ships that sank centuries ago have preserved records of the times in which they sailed. Until very recently when sophisticated instruments such as magnetometers, sonar, and remotely controlled video cameras have been developed, the most common source of oceanic discovery was the sponge diver of the Mediterranean Sea. Many times these men have found wrecked ships, or signs of them, while bringing up sponges from the sea floor at depths of over 100 feet.

As one shipwreck explorer, George Bass, wrote in the December 1987 *National Geographic*:

> Long experience has taught us that the best sources of information about ancient shipwrecks are the divers on Turkey's sponge boats. . . [they are] far more valuable than the most sophisticated sonar and magnetometers . . . In a single summer season the divers on 25 boats spend a combined total of about 20,000 hours roaming the seabed in the quest for sponges.

In a sense, using information provided by sponge divers as the basis for archaeological explorations makes ensuing discoveries less than serendipitous. But the first discoveries based on the observations by sponge divers were indeed serendipitous; the divers had no intention of finding ancient ships—they were interested only in finding sponges.

Bronze Age Ship off Turkey. Peter Throckmorton, a diver-explorer, met the burly Turkish captain of a sponge boat who had brought up an ancient bronze statue and a clay urn from the bottom of the Mediterranean Sea. The captain disdained the modern scuba equipment that the American group of diver-explorers used. He thought the equipment was for tourists, but not for real divers. However, the Americans and their equipment soon earned the respect of the Turkish divers. In fact, the younger sponge divers learned to use the scuba equipment in preference to their archaic metal diving helmets.

The summer of 1958 passed without any definite findings, but an idle mention by one of the sponge divers of corroded bronze ingots shaped like animal hides brought Throckmorton and his crew back the next summer, when they located the wreckage site of a Bronze Age ship. Although they could not find the actual ship, they recovered many bronze ingots, spear points, axes, and crude pottery items. Deteriorating weather delayed further search for the ship until the next summer, 1960. They finally found the ship, or at least fragments of its hull, and enough artifacts of its cargo to allow dating the last voyage of the ship at about 1,200 B.C. They placed more than ton of bronze and copper objects from the ship's cargo in a museum in Bodrum, Turkey.

Greek Ship off Cyprus. A Greek ship was recovered that sank off the coast of Cyprus in the fourth century B.C. Here, too, the key to the discovery was a sponge diver. He spotted a mass of amphorae, pottery wine jars used at that time, which were often cargo on merchant ships of the Mediterranean. In the June 1970 *National Geographic* Michael Katzev quotes the Cypriot diver telling how he found the jars and then had difficulty finding them again:

> I was diving for sponges when I suddenly noticed the anchor of my boat dragging. I followed it, and it slid right past that mound. But when I got to the surface, I was in a gale with no time to take bearings. For three years I tried to locate the jars again. You're lucky. It was only a few weeks ago that I got my second look at her. She's yours now. Only archaeologists must touch her. I've kept the secret for just such a group as yours, and to assure proper honor to my town. You must not forget that she's part of Kyrenia's history.

The wooden hull of the ship was virtually intact, preserved by the sand that had covered it for 22 centuries, as well as by a shell of lead that had been intended to protect the wood from shipworms. Apparently the ship's crew put too much faith in the lead shield; the archaeologists found that the worms had done extensive damage beneath the lead, and they speculated that the tunneling by the worms might have contributed to the sinking of the ship.

Bronze Age Ship at Ulu Burun. In 1973 George F. Bass founded the Institute of Nautical Archaeology (INA) at Texas A&M University, where he is Abell Professor of Nautical Archaeology. The INA, together with the National Geographic Society, the National Science Foundation, the National Endowment for the Humanities, and the Institute for Aegean Prehistory, is funding the excavation of the shipwreck at the Ulu Burun site, which began in 1984 and was still continuing in 1988.

The first hint of the wreck at Ulu Burun came in the summer of 1982. A young sponge diver told his captain that he had seen strange

▼ *The Bronze Age shipwreck at Ulu Burun, Turkey. The excavator uncovers rows of four-handled copper ingots about 160 feet deep.*

"metal biscuits with ears" on the sea floor while working at a depth of 150 feet off an underwater point. The captain recognized this description as that of a Bronze Age copper ingot from a drawing that INA had circulated among the sponge boat owners. He relayed the news to the INA organization. A preliminary survey was made in the summer of 1983; the photographs and sketches were such as to make Dr. Bass exclaim, "We're looking at an archaeologist's dream!" Thousands of items have been retrieved from the wreck, representative of the seven civilizations that flourished in the eastern Mediterranean in the late Bronze Age—the time of the reign of Tutankhamun in Egypt and the fall of Troy.

Among these items are hundreds of copper ingots, tin ingots (which were used with copper to make bronze), a golden chalice, a terracotta Mycenaean cup useful as a dating tool (indicative of the fourteenth century B.C.), a small gold scarab engraved with Queen Nefertiti's name, ebony from Africa of the same type used for furniture in Tutankhamun's tomb, Baltic amber beads, cobalt blue glass ingots perhaps from Canaan, amphorae filled with an aromatic resin used as incense, hundreds of mollusk shells of the type from which Tyrian purple (the purple dye of royalty) was extracted, and parts of a small wooden "book" with ivory hinges, probably representing the world's oldest book. The wooden "pages" of the book had been spread with a wax, which could be inscribed with a stylus. No wax remained, but one of the ingredients for the wax was found in a sealed amphora among the wreckage. The timbers of the boat's hull were fastened together by mortise-and-tenon joints pinned with hardwood pegs, in the same way Homer described the construction of the ship of Odysseus.

The continuing archaeological study has convinced the INA group that this ship was very special, perhaps carrying a cargo meant for royalty. Its shipwreck was likely a severe blow to many important persons of the fourteenth century B.C. However, the serendipitous discovery of its wreckage has provided an enormous amount of new information about that ancient era.

▼ POSTSCRIPT

Many other archaeological discoveries have been made, not only by "really trying," but also by clever design, or *conception*, as Sir Derek Barton would say. (See Chapter 35.) Among the many examples that might be cited is the finding of the Titanic. This achievement, which captured the attention of the western world, displayed the brilliant planning of many persons and the utilization of new scientific equipment, some of which was specifically designed for this project. It has been described and illustrated magnificently in two articles in the *National Geographic* in December 1985 and 1986, and in the book by Robert D. Ballard, *The Discovery of the Titanic* (1987).

19 ▼

Some Astronomical Serendipities

The Big Bang

▼ In 1964 Arno Penzias and Robert Wilson, scientists at Bell Laboratories in Holmdel, New Jersey, were modifying a radio antenna that had been used to receive signals from the early communication satellites. They intended to use the antenna for some rather prosaic studies on radio signals from outer space. In preparation, they tried to eliminate all terrestrial sources of background radio signals. They evicted a pair of pigeons that were nesting in the horn-shaped antenna, and removed what was referred to in elegant scientific prose as "a white dielectric substance." After taking all such precautions, they found that there was still a residual radiation "noise", comparable to static on a radio.

Astronomers had a theory that the universe started 15 billion years ago with a tremendous explosion of highly condensed matter—the Big Bang theory, as they euphemistically called it. This explosion produced a tremendous radiation of energy, which has been decreasing ever since. James Peebles at Princeton presented a paper about this theory at a scientific meeting at Johns Hopkins University in early 1965. In some way (various versions have been given), Penzias and Wilson at Bell Labs heard of Peebles' paper and of his theories about the Big Bang. When the Princeton and Bell groups exchanged information, they concluded that the "noise" detected by the Bell radio antenna had just the energy expected for the radiation left over from the Big Bang. "Either we've seen the birth of the universe, or (as one astrophysical folk tale would have it) we've seen a pile of pigeon ____!"

Apparently the Nobel Prize authorities accepted the more scientific version of this conclusion, for Penzias and Wilson were awarded the Prize in physics in 1978.

Pulsars

Jocelyn Bell and Anthony Hewish were not trying to discover pulsars in 1967. How could they? No one suspected that neutron stars emitted pulsed radio-frequency signals. At Cambridge University that summer, Bell and Hewish were trying to measure the size of radio sources by seeing whether the sources "twinkled" as their radio waves passed through the interplanetary medium.

But Bell noticed something unusual on the weekly extended charts that the Cambridge telescope produced: bursts of radiation appeared on the records each midnight. By the end of September she and Hewish had eliminated the possibility that these signals were terrestrial noise of any kind, and they noted that the bursts came earlier each night, just as stars do. When the signals became very strong in November, the astronomers detected pulses of short duration and at very regular intervals. Bell then searched through reams of chart records, and found three more pulsars. When these findings were announced, explanations were sought and many were given, ranging from facetious ("communications from 'Little Green Men' in outer space") to serious (some astronomical object somehow creates the pulses).

An answer came from David Staelin and Edward Reifenstein at the National Radio Astronomy Observatory in Green Bank, West Virginia, who found a pulsar in the center of the Crab Nebula. Pulsars were said to be neutron stars, the corpses left after a supernova explosion.

Pluto's Moon

An accident to a machine led to an astronomical discovery by James Christy at the U.S. Naval Observatory in 1978. Christy was measuring the orbital characteristic of Pluto. To do so, he had placed a photographic plate containing a picture of Pluto on an instrument called a Star Scan machine. When he did so, he noticed an elongation of the image of the planet. At first he assumed the bulge was an artifact and was going to discard the photograph. Luckily, however (as it turned out), the machine began to malfunction at that instant. Christy called in an electronics technician to repair the machine. The technician asked Christy to stand by while he made the repairs, because he thought he might need Christy's help.

During the hour required for the repair, Christy studied the photograph more carefully, and as a result he decided to look through the archives for earlier pictures of the planet. The first one he found was marked "Pluto image. Elongated. Plate no good. Reject." His interest now aroused, Christy searched through the archives and found six more pictures dated between 1965 and 1970 that showed the same bulge. His further studies proved that the bulge was a moon of the planet. If the Star Scan machine had not broken down when it did, he would not have discovered the new moon.

20 ▼

ACCIDENTAL Medical Discoveries

Insulin

▼ In 1889 in Strasbourg, Germany, while studying the function of the pancreas in digestion, Joseph von Mering and Oscar Minkowski removed the pancreas from a dog. One day thereafter a laboratory assistant called their attention to a swarm of flies around the urine from this dog. Curious about why the flies were attracted to the urine, they analyzed it and found it was loaded with sugar. Sugar in urine is a common sign of diabetes.

Von Mering and Minkowski realized that they were seeing for the first time evidence of the experimental production of diabetes in an animal. The fact that this animal had no pancreas suggested a relationship between that organ and diabetes. Von Mering and Minkowski subsequently proved that the pancreas produces a secretion that controls the use of sugar, and that lack of this secretion causes defects in sugar metabolism that are exhibited as symptoms of diabetes.

Many attempts were made to isolate the secretion, with little success until 1921. Two researchers, Frederick G. Banting, a young Canadian medical doctor, and Charles H. Best, a medical student, were working on the problem in the laboratory of Professor John J. R. MacLeod at the University of Toronto. They extracted the secretion from the pancreases of dogs. When they injected the extracts into dogs rendered diabetic by removal of their pancreases, the blood sugar levels of these dogs returned to normal or below and the urine became sugar-free. The general condition of the dogs also improved.

Professor MacLeod took an active interest in the project, improving procedures for extracting the hormone and standardizing its dosage. He also suggested the name *insulin* for the secretion, which is now recognized as a hormone, when it was found to be produced by clumps of isolated, or insular, cells in the pancreas called the islet of Langerhans. Within a year, purified extracts of beef pancreases were tested on human diabetics and were found to relieve the symptoms of the disease. One of the first human subjects was a volunteer, desperately ill with diabetes, a classmate and friend of Banting. His condition improved dramatically and he served as a human guinea pig for standardization tests.

Before 1922 (when insulin was first used clinically), dietary treatment of diabetes had helped to minimize symptoms and prolong life, but it was far from satisfactory. Diabetes was a severe, debilitating disease, ultimately fatal in most cases. Infections of various kinds were very likely to occur, surgery was hazardous, and childbearing was dangerous to both mother and child. Insulin can control most cases of diabetes, diminish or remove the incidental dangers, and prolong useful activity and life.

Banting and MacLeod announced their isolation and clinical use of insulin against diabetes in a paper read before the American Physiological Society in January 1922. They shared the Nobel Prize in Physiology or Medicine in 1923. The remarkably short time between the discovery and its honor by the prize testifies to its importance in medical practice.

Insulin is a protein, a natural polymer made up of 51 amino acids joined, in a specific sequence, in two connected rings. The sequence of the amino acids in the rings of bovine insulin was determined in 1953 by Frederick Sanger of Cambridge University, who received the Nobel Prize in Chemistry for this accomplishment in 1958. The sequence of amino acids is slightly different in certain species of animals, but the differences in sequence are not critical to the regulating effect on carbohydrate metabolism in humans; someone who becomes allergic to equine insulin, for example, can switch to porcine insulin. Until recently all insulin used for the treatment of human diabetes came from the pancreases of horses and pigs. As a result of genetic engineering, based on knowing how DNA controls protein synthesis, a major pharmaceutical firm has begun to produce human insulin by using a bacterium, *escherichia coli.* This process promises to make available an abundant supply of insulin that does not depend on sometimes unreliable animal sources.

Von Mering and Minkowski are remembered now not for their contributions to digestion, but mainly for their pioneering work on the cause and control of diabetes. Banting and Best have received more credit for their contributions to the treatment of this terrible disease, but the initial insight into its cause was provided by the alert response of von Mering and Minkowski to the serendipitous sighting of flies swarming over the urine from a depancreatized dog, which might well have gone unnoticed as a bothersome but trivial incident.

Allergy, Anaphylaxis, and Antihistamines

Charles Robert Richet received the Nobel Prize in Physiology or Medicine in 1913 for his discoveries related to allergies and anaphylaxis.

Richet was born in Paris in 1850, the son of a professor of clinical surgery. He, in turn, became a professor of physiology, but he did not limit his interest to that discipline. He published papers on physiological chemistry, pathology, pharmacy, and psychology. He followed the development of aeronautics closely and actually designed an aircraft. He was active in pacifist movements. His last two interests are somewhat poignant in that one of his sons was killed as a pilot in World War I.

His prize-winning work grew out of a totally unexpected result. At the Nobel Prize ceremony, Richet described his discovery in modest terms:

> Let me tell you under what circumstances I observed this phenomenon for the first time. I may be permitted to enter into some details on its origin. You will see, as a matter of fact, that it is not at all the result of deep thinking, but of a simple observation, almost accidental; so that I have had no other merit than that of not refusing to see the facts which presented themselves before me, completely evident.

Richet went on to describe a cruise he made on the yacht of Prince Albert of Monaco, and how the Prince had encouraged him to study the poison of a sea animal called the Portuguese man-of-war. (Many have encountered these unwelcome visitors to beaches and have suffered excruciating pain from their stings.) When Richet returned to France, where he was professor of physiology at the University of Paris, he could not conveniently obtain this particular animal and decided to study instead the poison from the tentacles of Actiniae, sea anemones that are common on the rocks along European coasts. He extracted the poison into glycerine and proceeded to determine the toxic dose, using dogs as test animals. These tests were not simple because the poison took effect slowly, sometimes requiring several days to exert its maximum effect.

Some dogs escaped death, perhaps because they had received a less-than-lethal dose or for some unknown reason. After several weeks, when these dogs appeared to have been restored to normal health, they were used again in new experiments. Then, something extraordinary and unforeseen occurred. Dogs that had received a first dose of poison and had survived, upon being given a *much weaker* dose the second time, immediately showed dreadful symptoms: vomiting, loss of consciousness, asphyxia, and death. Richet continued his account:

> Repeating on different occasions this fundamental experiment, we were able to establish, in 1902, these three principal facts, which are the very foundation of the story of anaphylaxis: first, an animal injected beforehand is enormously more sensitive than a new animal; second, the symptoms which supervene on the second injection, characterized by a rapid and total depression of the nervous system, have no resemblance to the symptoms produced by the first injection; third, this anaphylactic state requires an interval of three or four weeks to establish itself. It is what is called an incubation period.

Richet and others proceeded by their research to generalize the new phenomenon of anaphylaxis, or sensitization, of the body to injections of minuscule quantities of proteins. Richet demonstrated the transmission of the effect by taking the blood of a sensitized animal and injecting it into a normal animal to produce the anaphylactic state in the second animal. Thus the identity of the agent was shown to be a chemical substance in the blood.

Richet coined the term *anaphylaxis* to describe a state of sensitization that is the antithesis of prophylaxis, a condition of immunity produced by injections of bacterial toxins, for example, a subject that Richet had also studied. The synonym *allergy* is now more commonly used to describe this sensitization.

Diseases of allergy now form a major division of internal medicine. Two main forms of treatment are used for allergies. One is desensitization when the antigen or antigens have been identified; this is done by introducing minute amounts of the suspected agents causing the allergic reactions at intervals, and gradually increasing the doses until resistance has built up in the subject's body. The second form of treatment is the use of an antihistaminic drug—one that offsets the effect of the histamine released in the subject's body by the agents to which the subject is allergic. Antihistaminic drugs are familiar to everyone. Both prescription and nonprescription types account for a large fraction of the pharmaceutical market. This field of medical science and practice started with the investigations of Charles Richet.

These were initiated by an unexpected reaction of a dog to a tiny dose of poison from a sea anemone—but Richet recognized the extraordinary nature of this reaction and found an explanation for it.

Nitrogen Mustards and Cancer Chemotherapy

An accidental exposure of troops to mustard gas during World War II marked a turning point in the approach to cancer chemotherapy. Mustard gas is actually a liquid. The combatants never intentionally used it in World War II, but both sides kept it near the war fronts for retaliation if the other side used it first. It was called a gas because, if used, it would be dispersed by exploding shells, which would vaporize it and spread it on the surrounding areas, including humans.

An Allied ship carrying mustard gas was bombed in an Italian harbor and after the poisonous agent had spread on the water, some combat personnel were thrown into the water. After they were rescued, they were treated for the effects of the mustard gas. Many of these patients developed a blood disorder manifested by a dangerous reduction of the white blood cells. Because a decrease in white cell count would indicate improvement in the case of certain leukemias in which leucocytes are overproduced in the neoplastic bone marrow, the mustard gas was subsequently tested in leukemia patients. The high toxicity of the agent was prohibitive, but because of the similarity in molecular structure of *nitrogen mustards* to sulfur mustards they were tested against blood cancers. A nitrogen mustard is a compound similar to the most common (sulfur) mustard gas, but in the nitrogen mustard a nitrogen atom takes the place of the sulfur atom in the sulfur mustard molecule.

Many variations of both nitrogen and sulfur mustards were prepared as potential war gases. After the bombing that resulted in the accidental poisoning of our troops, several hundreds of analogous nitrogen mustards and other structurally related compounds were tested as anticancer agents. Although none cured any type of human cancer, their tumor-retarding effects raised the hope that curative compounds might be found in the long run.

The Pill

The average medicine cabinet and the shelves of the local pharmacy contain many kinds of pills, but few people doubt which one is meant when someone refers simply to the Pill. The oral contraceptive introduced in the 1960s had an enormous impact on our culture. It has been said to be partly responsible for the sexual revolution, the emancipation of women, and the secularization of the Catholic church in the Western world. The origin of the Pill is not one serendipitous event; it is more complicated than that, but serendipity is certainly one part of the story.

The story begins with an archetypical entrepreneur, Russell E. Marker. While a member of the Pennsylvania State University chemistry faculty in the late 1930s, Marker discovered a way to make the female sex hormone progesterone from a fairly common type of steroid called sapogenins. Progesterone was valuable for treating menstrual disorders and for preventing miscarriages. Until this time, this drug was available only from European pharmaceutical companies where it was made by laborious syntheses, and it was extremely expensive.

Marker learned that sapogenins of the type he needed were abundant in some types of yams that grew wild in Mexico. He was unable to get support from his university or any American pharmaceutical firm to

▼ ***Russell Marker and Mexican yams***

finance an expedition to acquire the yams and try his procedure to make progesterone from them. So he resigned his position at Penn State, went to Mexico, rented a cottage, and set out by mule to the jungle-covered hills of southern Mexico. He collected 10 tons of yams and isolated the sapogenin he wanted, diosgenin, working in a rented laboratory in Mexico City. Back in the United States in a friend's lab, he used his process to synthesize 2,000 grams of progesterone, worth at that time about $160,000.

Returning to Mexico City, he located a small laboratory through the telephone directory and convinced the two owners to join him in a gamble to manufacture progesterone. They named their company Syntex. Marker stayed with Syntex less than two years because of disputes with his associates, but the Mexican partners in Syntex recruited a Swiss-trained chemist from Cuba, George Rosenkranz, who was able to continue the synthetic work and even succeeded in preparing the male sex hormone testosterone from the same Mexican yams. Syntex broke the European cartel and the price of the hormones dropped from $80 to about $1 per gram.

In 1949, Carl Djerassi was asked to head a research team at Syntex for the production of another steroid, cortisone, which had been discovered by E. C. Kendall a few years earlier, and was considered a wonder drug by this time. Djerassi was born in Vienna, Austria, emigrated to the United States, earned an A.B. degree from Kenyon College at age 19 and a Ph.D. from the University of Wisconsin when 22. He was a research chemist at Ciba Pharmaceutical Co. in New Jersey when he was asked to go to Mexico to direct the steroid research for Syntex.

Djerassi said that when he went to Syntex, "an oral contraceptive was not in anyone's plans." The main objective was cortisone, but another research target was estradiol, a female hormone used to treat certain disorders that can occur at puberty and menopause. While trying to synthesize a molecule having the biological action of estradiol, Djerassi and his chemists at Syntex accidentally made one that was more like progesterone. This compound, 19-norprogesterone, contained just one less carbon atom in the molecule than progesterone. It was found to be more potent in the body than the natural hormone, but it could be given only by direct injection into the blood stream.

Progesterone has sometimes been referred to as nature's contraceptive, since one of its functions is to inhibit ovulation during pregnancy. The Syntex chemists' next aim was to modify the synthetic progesterone-type compound, 19-norprogesterone, so that it could be taken by mouth rather than by injection. They were able to do this by a slight chemical modification, taking a hint from research published by a German chemist, Hans Inhoffen, a dozen years earlier. The modified compound (norethindrone) had the potent progesterone-like activity desired, and it was also stable in the stomach, and so it could be taken by mouth.

Thus was born the first oral contraceptive—it started with Russell Marker's discovery of a new way to convert sapogenins to sex hormones, his bold venture in the Mexican tropical jungles to obtain sapogenins from wild Mexican yams, the gamble of a tiny new laboratory to challenge the dominant European cartel. It continued with Carl Djerassi's accidental production of a synthetic contraceptive much like progesterone, and his intentional modification of it so that it could be taken orally.

Demonstrating that it could be done soon led others to find more ways to do it, and now there are several synthetic oral contraceptives. Nevertheless, the compound first synthesized in the small Mexican pharmaceutical laboratory, norethindrone, is still the active ingredient in almost half of all oral contraceptives.

▼ POSTSCRIPT

Russell Marker not only resigned from Syntex, he resigned from chemistry. After 1949, when he was 47, he avoided chemical research and production. He had been exceedingly independent from an early age. The story goes that he never received a degree in chemistry from the University of Maryland because he refused to take physical chemistry, a required subject. In spite of this he attained a research position in a major university and made chemical history in a short early period of his life.

After three years in Mexico, Carl Djerassi took an appointment at Wayne State University, where he rose from associate professor to professor by 1959. He retained a connection with Syntex in research and with Zoecon Corporation later. In 1959 he went to Stanford University as Professor of Chemistry. He has been given numerous honorary degrees and awards for his pioneering research in antihistamines and oral contraceptives.

LSD

The discovery of the hallucinogenic substance lysergic acid diethylamide (LSD) is one of the most frightening accounts in recorded medical history.

LSD is derived from lysergic acid, which itself is not hallucinogenic. Lysergic acid occurs along with a large number of poisonous alkaloids in the fungus ergot, which sometimes forms on rye in the field during a particularly wet season. For centuries this fungus plagued persons from Spain to Russia who, out of ignorance or hunger, ate bread baked from contaminated rye flour. Gangrene of extremities resulted from extreme constriction of blood vessels. The affliction was called St. Anthony's fire because the victims had the terrible sensation that their skin was burning, their blackened fingers and toes look charred, and they sought relief at St. Anthony's shrine. During the Middle Ages eating spoiled rye flour reportedly caused abortions, visual disturbances, and mental aberrations culminating in epidemics of madness. These symptoms were probably caused by overdoses of the poisonous ergot alkaloids combined with hysteria engendered by the mutilating affliction rather than by lysergic acid itself. Not until Albert Hofmann, a Swiss chemist employed by the Sandoz Laboratories in Basel, synthetically attached the diethylamide group did the acid acquire its mind-bending properties. Hofmann was studying lysergic acid and related compounds in the hope of developing a drug to treat migraine headaches or to control bleeding after childbirth.

The following account of his experience in this research is taken from Hofmann's own notebook as reported in *The Beyond Within: The LSD Story* (1970) by Dr. Sidney Cohen:

> Last Friday, the 16th of April [1938], I had to leave my work in the laboratory and go home because I felt strangely restless and dizzy. Once there, I lay down and sank into a not unpleasant delirium which was marked by an extreme degree of fantasy. In a sort of trance with closed eyes (I found the daylight unpleasantly glaring) fantastic visions of extraordinary vividness accompanied by a kaleidoscopic play of intense coloration continuously swirled around me. After two hours this condition subsided.

Hofmann suspected that his unusual sensations might be due to accidentally swallowing or inhaling a minute amount of some chemical in the laboratory. His account continues:

> On that Friday, however, the only unusual substances with which I had been in contact were D-lysergic acid and isolysergic acid diethylamide. I had been trying various methods of purifying these isomers by condensation, and also breaking them down into their components. In a preliminary experiment I had succeeded in producing a few milligrams of lysergic acid diethylamide (LSD) as an easily soluble crystal in the form of a neutral tartrate [a salt formed from LSD and tartaric acid]. It was inconceivable to

> me, however, that I could have absorbed enough of this material to produce the above described state. Furthermore, the symptoms themselves did not appear to be related to those in the ergotamine-ergonovine group. I was determined to probe the situation and I decided to experiment upon myself with the crystalline lysergic acid diethylamide. If this material were really the cause, it must be active in minute amounts, and I decided to begin with an extremely small quantity which would still produce some action in equivalent amounts of ergotamine or ergonovine.

Hofmann therefore took 250 micrograms [0.00025 gram] of lysergic acid diethylamide. After 40 minutes he noted a "mild dizziness, restlessness, inability to concentrate, visual disturbance, and uncontrollable laughter." At this point the entries in the laboratory notebook end, and the last words were written only with the greatest difficulty. He continued his account later:

> I asked my laboratory assistant to escort me home since I assumed that the situation would progress in a manner similar to last Friday. But on the way home (a four-mile trip by bicycle, no other vehicle being available because of the war), the symptoms developed with a much greater intensity than the first time. I had the greatest difficulty speaking coherently and my field of vision fluctuated and was distorted like the reflections in an amusement park mirror. I also had had the impression that I was hardly moving, yet later my assistant told me that I was pedaling at a fast pace.
>
> So far as I can recollect, the height of the crisis had passed by the time the doctor arrived; it was characterized by these symptoms: dizziness, visual distortions, the faces of those present appeared like grotesque colored masks, strong agitation alternating with paresis [partial paralysis], the head, body and extremities sometimes cold and numb; a metallic taste on the tongue; throat dry and shriveled; a feeling of suffocation; confusion alternating with a clear appreciation of the situation; at times standing outside myself as a neutral observer and hearing myself muttering jargon or screaming half madly.
>
> The doctor found a somewhat weak pulse, but in general a normal circulation. Six hours after taking the drug, my condition had improved definitely.
>
> The perceptual distortions were still present. Everything seemed to undulate and their proportions were distorted like the reflections on a choppy water surface. Everything was changing with unpleasant, predominantly poisonous green and blue color tones. With closed eyes multihued metamorphizing fantastic images overwhelmed me. Especially noteworthy was the fact that sounds were transposed into visual sensations so that from

each tone or noise a comparable colored picture was evoked, changing in form and color kaleidoscopically.

After a good night's rest, Hofmann felt "completely well, but tired." The accidental ingestion of LSD by this perceptive chemist initiated a chain of investigation of chemically induced mental alterations that has extended into every psychiatric research center. The importance of Hofmann's discovery is not that LSD has any direct chemical relationship to a disease such as schizophrenia; its structure can hardly be expected to be synthesized by the human metabolism.

The discovery of LSD has other significant implications; it demonstrates that chemical substances in extremely minute amounts can induce mental distortions that resemble the naturally occurring psychoses. It has stimulated interest in the chemistry of the nervous system, especially the chemical transmitters across synapses, the nerve-cell connections; and it permits the laboratory study of both normal and abnormal mental processes.

Although many lysergic acid derivatives with hallucinogenic activity have been synthesized, none is as potent as LSD. Completely new and unrelated chemical groups have also been found to possess similar psychic properties. One day the chemistry of mental illness may be clear, in part, thanks to the serendipitous and terrifying experience of Albert Hofmann.

▼ POSTSCRIPT

Because LSD was abused with such serious consequences, the Sandoz Laboratories discontinued production of the drug in 1966 and turned over all their supplies to the National Institute of Mental Health.

In a later chapter in his book, entitled "The Worst That Can Happen", Dr. Cohen comments on the possible use and abuse of LSD: "The title of this chapter is a phrase borrowed from the siren song of the LSD Religion spokesmen. 'The worst that can happen to you after taking an LSD trip' they continue to assert 'is that you will come back no better than you were.' This is as incorrect as a statement can be. You may come back much worse than you were. You may not come back."

The Pap Test

"Twenty-five years ago uterine cancer was the biggest killer of American women; today, according to the American Cancer Society, 180,000 women are 'cured, alive and well' five years after treatment, primarily as a result of the smear test." This quotation is from a report of the death of Dr. George Nicholas Papanicolaou in 1962 in *Medical World News*.

Dr. Papanicolaou developed the Pap smear test after a serendipitous observation he made in 1923. He first reported the observation in 1928 in a brief paper titled "New Cancer Diagnosis" that he gave at a conference in Battle Creek, Michigan. The paper was subsequently recorded in a journal called *Growth*. The medical profession paid little attention to this report, however, until it was noticed by Dr. Joseph C. Hinsey, who in 1940 was dean at Cornell University Medical College. Dr. Hinsey encouraged "Dr. Pap" to resume his research on uterine cancer in a laboratory put at his disposal at Cornell. In 1943 a paper with Dr. Herbert F. Traut titled "Diagnosis of Uterine Cancer by the Vaginal Smear" finally aroused the interest of the medical community. In 1948 Dr. Charles S. Cameron, Medical and Scientific Director of the American Cancer Society, called a conference in Boston. Thereafter, the Pap test was gradually adopted worldwide.

George Papanicolaou's remarkable life began in 1883 in the small town of Coumi, Greece. He was the son of a doctor, and obtained his medical degree at the University of Athens. Deciding to devote his life to research rather than to practice as a family doctor, he went to Germany and studied at Jena, Freiburg, and Munich, obtaining a Ph.D. at the University of Munich in 1910. His professional career was interrupted by service in the Greek army in the Balkan War, where he met Americans of Greek ancestry whose descriptions of America as a land of opportunity persuaded him to seek his future there. He came to America with his wife, whom he had married upon his return to Greece after completing his study at Munich, with no definite plan about how to fulfill his ambitions for research in biology and medicine.

He spent one day as a rug salesman in a New York department store. On recommendation of Dr. Thomas Morgan, professor of Zoology at Columbia University, who was acquainted with his research at Munich, he was given a part-time position in the department of pathology in a New York hospital that was affiliated with Cornell University Medical College. Soon he was transferred to the Department of Anatomy of the Medical College, where he progressed to assistant professor, then professor, professor emeritus, director of the Papanicolaou Research Laboratory, and consultant to the Papanicolaou Cytology Laboratory.

His association with Cornell spanned nearly half a century. He worked 14 hours a day, six and a half days a week in his Cornell laboratories and at home, chiefly assisted by his wife. He took only one vacation in 41 years, explaining when asked, "Work is too interesting and there is so much to be done!"

The recognition and success Dr. Papanicolaou achieved followed an observation that he, himself, declared to be "largely a matter of luck." In 1917, Professor Charles R. Stockard, the chairman of the Anatomy Department at Cornell University Medical College, invited Dr. Pap to join him in studies of experimental genetics. At this time there was great interest in the role of chromosomes in the determination of sex. Dr. Pap began his work with guinea pigs; this research led to a paper with Dr. Stockard in 1917 that described the cellular changes observable in the vaginal tissue during different stages of the estrus cycle of the guinea pig. In order to learn if comparable vaginal cellular changes were characteristic of the menstrual cycle of the human female, Dr. Pap undertook in 1923 a systematic study of the cytology (cell biology) of human vaginal fluid; this work was carried out on patients in the Woman's Hospital of New York.

A specimen was obtained from a woman with cancer of the uterus. Because of Dr. Pap's faculty for acute observation and sense of importance, he recognized the structural abnormalities of cancer cells in the smears of the vaginal fluid. He said later, "The first observation of cancer cells in the smear of the uterine cervix gave me one of the greatest thrills I ever experienced during my scientific career." At the end of the 1928 paper, after describing the new test for preinvasive uterine cancer, he predicted: "A better understanding and more accurate analysis of the cancer problem is bound to result from the use of this method. It is possible that analogous methods will be developed for the recognition of cancer in other organs. I feel that such methods can and will be developed in the future." Dr. Pap fulfilled his own prediction. His method came in time to be applied to the colon, kidney, urinary bladder, prostate, lung, stomach, breast, sinuses, and even to the brain.

▼ POSTSCRIPT

The honors that Dr. Papanicolaou received are too numerous to list here. They include having a cancer research institute named for him, several awards from cancer societies and women's societies, three honorary university degrees, and two awards from the Greek government.

The Pap test for preinvasive cervical cancer has been recognized as one of the greatest medical triumphs of life saving. In 1959 the director of the Institute of Oncology in Moscow reported that eight million women had been screened by the Pap test in Russia. In Sweden, where the most complete medical records in the world are kept, 207,455 women were tested over a ten-year period, and records were computerized. The results showed that there was a 75% decrease in the incidence of cervical cancer among women who had smears taken at least once during the ten-year period. Doctors at the University of Uppsala estimate that a system in which a woman undergoes one Pap smear every three years can reduce the incidence of invasive cervical cancer to a level between one and five cases per 100,000 women per year.

Light and Infant Jaundice

Niels Filsen was given the Nobel Prize in Physiology or Medicine in 1903 "in recognition of his contribution to the treatment of diseases, especially lupus vulgaris, with concentrated light rays." The value of sunlight in the prevention of the bone disease rickets has been known since 1919; the light converts the precursors of vitamin D in the body into the effective vitamin, which controls the calcification of bones.

More recently several other beneficial effects of light have been claimed, some of which have been discovered by accident. And some of the claims of the efficacy of light, both natural and artificial, are controversial. One that is accepted and used by the medical profession, and was discovered serendipitously, is the effect of sunlight on infant jaundice.

Newborn babies sometimes show a yellowing of their skin known as infant jaundice. The cause of the yellow tint is the same as in other forms of jaundice—the bile pigment bilirubin is present in an abnormal concentration and shows up in the skin and in the whites of the eyes. Bilirubin is a degradation product of hemoglobin; its abnormal high concentration in the blood signals a malfunction of the liver, spleen, or gallbladder. In the case of infant jaundice, the baby's immature liver cannot remove the bilirubin fast enough to prevent brain damage and death.

In the late 1950s a perceptive nurse in a hospital in England noticed that when jaundiced babies were placed near the nursery windows in the sunshine, their jaundice faded. Research initiated after the nurse's report of this observation demonstrated that the ultraviolet component of the sunlight bathing the all-but-transparent skin of the babies converted the bilirubin into an excretable form. Irradiation of infants with ultraviolet light has now become standard hospital practice to prevent or cure infant jaundice.

Cholesterol Receptors

Deposition of cholesterol in blood vessels leading to heart failure is one of the major health concerns of our time. Drs. Michael S. Brown and Joseph L. Goldstein, of the University of Texas Health Science Center at Dallas, received the Nobel Prize in Physiology or Medicine in 1985 for their discovery of cholesterol receptors vital to the reduction of cholesterol in the bloodstream. Work in this area has been carefully planned basic scientific studies by brilliant, highly trained, and dedicated researchers, but even so it has benefited from serendipity.

"In the beginning, we started with a hypothesis that was incorrect," Brown recalled. "We were studying familial hypercholesterolemia (FH), a disease in which . . . children had extremely high levels of cholesterol in their blood. Initially we thought that an enzyme had gone wild and was producing the excessive cholesterol. As it turned out, we found the enzyme was not the problem. It was the fact that the [body] cells had trouble getting cholesterol from a lipoprotein" [and thus removing it]. "We did not dream there was any such thing as a receptor. It was not within the world view of any scientist that such a thing existed."

Brown and Goldstein began their study with cultured skin cells from FH patients, because it was difficult or impossible to obtain livers from such patients. Chance aided the research when Brown had a phone call from a doctor in Denver who had a liver from a patient with FH (obtained in a liver transplant operation). Since the skin cell studies were already underway, Brown asked for and received some skin from the patient rather than the liver.

As a result of their skin cell studies, Brown and Goldstein learned that FH patients lacked functional cell surface receptors for low density lipoproteins (LDL), the fundamental carriers of blood cholesterol to body cells.

At about the same time in 1973 as the Dallas workers were trying to understand how a lack of LDL receptors and a consequent excess of LDL in the blood can cause heart disease, a veterinarian in Japan made a serendipitous discovery that greatly aided the LDL receptor research. Yoshio Watanabe of Kobe University noticed that a rabbit in his colony had ten times the normal concentration of cholesterol in its blood. By appropriate breeding, Watanabe obtained a strain of rabbits, all of which had this high concentration of cholesterol in their blood, and all of these animals developed a coronary heart disease that closely resembled human heart disease. The rabbits were also found to lack the functional LDL receptors as in the case of humans with FH disease.

FH disease accounts for only a small fraction of coronary heart disease cases. What is still not understood is why some humans have high cholesterol levels even though they do not have FH disease and do not ingest large amounts of cholesterol. There is certainly a genetic factor that accounts for the fact that some individuals are more prone to have this problem than others. The studies of cell receptors in both humans and rabbits have given some answers to this question and promise to give more. Basically, high cholesterol concentration in the blood stream is due to two factors: one is the overproduction of cholesterol by the liver, the other is the underreception of cholesterol by the liver and adrenal glands, which produce vital steroidal hormones from cholesterol.

Fortunately there now seems to be a possibility of correcting this imbalance. A drug called cholestyramine has been found to increase the number of LDL receptors and a new drug isolated from a mold suppresses cholesterol synthesis in the liver. The latter drug, mevinolin (or lovastatin) was discovered independently by Akiro Endo of the Sankyo Drug Co. and by scientists at Merck Sharp and Dohme. Merck will market it as Mevacor. Brown and Goldstein have suggested that the "two-drug method" of lowering cholesterol levels may be worth trying for persons with critically high cholesterol concentrations.

21

X RAYS, Radioactivity, and Nuclear Fission

Discovery of X rays by Röntgen

In 1895 the German physicist Wilhelm Conrad Röntgen discovered X rays by accident. Röntgen was repeating experiments by other physicists in which electricity at high voltage was discharged through air or other gases in a partially evacuated glass tube. As early as 1858 it was found that the walls of the glass tube became phosphorescent during the discharge. In 1878 Sir William Crookes described the "cathode rays" causing this phosphorescence as a "stream of molecules in flight," but we now know that cathode rays are actually streams of electrons being emitted from the cathode, and the impact of these electrons on the walls of the glass tubes produces the phosphorescence.

Neon signs, television tubes, and fluorescent light tubes are modern developments of these experiments. The insides of fluorescent tubes are coated with highly fluorescent materials to produce different colors and shades of light.

In 1892 Heinrich Hertz demonstrated that cathode rays can penetrate thin metallic foils. Two years later Philipp Lenard constructed discharge tubes having thin aluminum windows. These windows allowed the cathode rays to pass out of the tube where they could be detected by the light they produced on a screen of phosphorescent material (such screens were also used to detect ultraviolet light); but they were found to travel only two or three centimeters in the air at ordinary pressure outside the evacuated tube.

Röntgen repeated some of these experiments to familiarize himself with the techniques. He then decided to see whether he could detect cathode rays issuing from an evacuated all-glass tube such as Crookes had used—that is, one with no thin aluminum window. No one had observed cathode rays under these conditions. Röntgen thought the reason for the failure might be that the strong phosphorescence of the cathode tube obscured the weak fluorescence of the detecting screen. To test this theory, he devised a black cardboard cover for the cathode tube. To determine the effectiveness of the shield, he then darkened the room and turned on the high voltage coil to energize the tube. Satisfied that his black shield did indeed cover the tube and allowed no phosphorescent light to escape, he was about to shut off the coil and turn on the room lights so that he could position the phosphorescent screen at varying short distances from the vacuum tube.

Just at that moment, however, he noticed a weak light shimmering from a point in the dark room more than a yard from the vacuum tube. At first he thought there must be, after all, a light leak from the black mask around the tube, which was being reflected from a mirror in the room. There was no mirror, however. When he passed another series of charges through the cathode tube, he saw the light appear in the same location again, looking like faint green clouds moving in synchronism with the fluctuating discharges of the cathode tube. Hurriedly lighting a match, Röntgen found to his amazement that the source of the mysterious light was the little fluorescent screen that he had planned to use as a detector near the blinded cathode tube—but it was lying on the bench more than a yard from the tube.

Röntgen realized immediately that he had encountered an entirely new phenomenon. These were not cathode rays that lit up the fluorescent screen more than a yard from the tube! With feverish activity he devoted himself single-mindedly in the next several weeks to exploring this new form of radiation. He reported his findings in a paper published in Wurzburg, dated December 28, 1895 and titled "A New Kind of Ray, a Preliminary Communication." Although he described accurately most of the basic qualitative properties of the new rays in this paper, his acknowledgement that he did not yet fully understand them was indicated by the

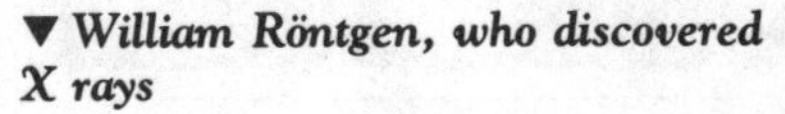

▼ ***William Röntgen, who discovered X rays***

name he chose for them, X rays. (They have often been called Röntgen rays.)

He reported that the new rays were not affected by a magnet, as cathode rays were known to be. Not only would they penetrate more than a yard of air, in contrast to the two- or three-inch limit of cathode rays, but (to quote his paper):

> All bodies are transparent to this agent, though in very different degrees. . . . Paper is very transparent; behind a bound book of about one thousand pages I saw the fluorescent screen light up brightly . . . In the same way the fluorescence appeared behind a double pack of cards . . . Thick blocks of wood are also transparent, pine boards two or three centimeters thick absorbing only slightly. A plate of aluminum about fifteen millimeters thick, though it enfeebled the action seriously, did not cause the fluorescence to disappear entirely . . . If the hand be held between the discharge tube and the screen, the darker shadow of the bones is seen within the slightly dark shadow image of the hand itself.

He found that he could even record such skeletal images on a photographic film. This property of X rays captured the attention of the medical world immediately. In an incredibly short time X rays were used routinely for diagnosis in hospitals throughout the world.

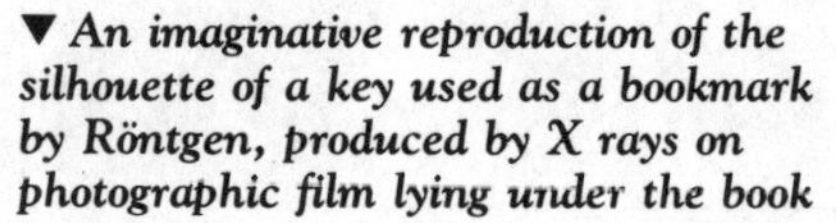

▼ ***An imaginative reproduction of the silhouette of a key used as a bookmark by Röntgen, produced by X rays on photographic film lying under the book***

Few events in the history of science have had so powerful an impact as Röntgen's discovery. Within a year after his first announcement, 49 books and pamphlets and over a thousand articles on X rays appeared. However, it was nearly 20 years before any significant advance was noted in the knowledge of the properties of X radiation over what Röntgen had obtained.

When the Swedish Academy of Sciences distributed the Nobel Prizes for the first time in 1901, the choice for the prize in physics was Röntgen. It certainly must have been a great satisfaction to the Academy to have such an eminent achievement to honor with the first award.

▼ POSTSCRIPT

Wilhelm Konrad Röntgen was born in Lennep, Prussia, in 1845. He received his early education in the Netherlands after his family moved there when Röntgen was three years old. Following brief study in Utrecht at the Technical School and at the University, he was admitted to the Polytechnical School in Zürich, where he received a diploma as a mechanical engineer. He became more interested in the pure sciences than in engineering, however, and began studies in mathematics and physics.

After studying with August Kundt, he was awarded the Ph.D. by the University of Zürich, with a thesis on "Studies of Gases." He followed Kundt to Würzburg a year later, then to Strasbourg, where Röntgen received his first appointment as a lecturer. In 1888 he accepted the post of professor of physics and director of the Physical Institute at the University of Würzburg. He held this post 12 years and here he discovered the X rays. In 1900 he was called by the government of Bavaria to head the Physical Institute at Munich, where he remained for the rest of his career. He died in 1923, at age 78.

Röntgen's life might have been severely shortened by the radiation he discovered had he not been shielded from the X rays during much of his experimentation. He had built a booth in his laboratory, not for health protection, but for the convenience of photographic development during the day. The lethal effects of overexposure to X rays were not fully appreciated during Röntgen's lifetime.

Discovery of Radioactivity by Becquerel

The discovery of natural radioactivity by Henri Becquerel followed quickly that of X rays by Röntgen, and for good reason. Becquerel read the paper in which Röntgen described his new penetrating rays as being produced by cathode rays, which also produced phosphorescence in the glass of the cathode tubes. Becquerel reasoned that certain substances made phosphorescent by visible light might emit a penetrating radiation similar to X rays—an incorrect theory, but one that led, regardless, to a valuable discovery.

Becquerel chose a phosphorescent compound of uranium. To test his theory, he wrapped a photographic plate in black paper, placed a crystal of the uranium compound on the paper-wrapped plate, and put the assembly in bright sunlight. When the photographic plate was developed, it bore an image of the uranium crystal. Becquerel, a careful experimenter, had previously determined that the black paper would protect the photographic plate from sunlight, and so he was certain that it was not sunlight alone that had exposed the plate. He considered the experiment a confirmation of his theory.

Then occurred the accident, or at least the intervention of a natural event, that led to a new era not only in chemistry and physics, but in life for all on this planet: the atomic and nuclear age. The sun did not shine in Paris for several days (a common occurrence). Because Becquerel considered the sunlight necessary to activate the phosphorescence of the uranium crystal, he suspended his experiments and put the uranium crystal away in a drawer, on top of a securely wrapped photographic plate.

After several days, Becquerel developed the photographic plate that had been in the drawer with the uranium crystal. He expected to find only a feeble image of the crystal, resulting from a small amount of residual phosphorescence in the uranium crystal. Instead, to his surprise, the image of the film was as strong as when the uranium crystal and wrapped film had been in sunlight! At this point, Becquerel drew the correct conclusion: the effect of the sunlight in producing phosphorescence of the uranium crystal had nothing to do with the exposure of the covered photographic plate beneath it, but this exposure came from the uranium crystal itself, even in the dark.

Becquerel began testing all of the samples containing uranium that he could find for the rays that exposed photographic film through black paper—rays that were obviously not ordinary light rays. He found that every pure uranium compound or even impure uranium ore had this property. He could measure the radiations from these materials by using an electroscope because the radiations ionized the air through which they passed. The operation of an electroscope is based on the fact that like charges repel. The force of repulsion is made observable by the deflection of a flexible conductor working against a mechanical restoring force.

Becquerel found the degree of radiation in all of the samples but one to be directly proportional to the percentage of uranium in the compound or ore. The one exception was an ore called pitchblende, which showed a radiation several times greater than that of pure uranium. This finding led Becquerel to conclude that this ore contained something besides uranium that had a much higher radioactivity than uranium.

At this point the Curies enter the story of radioactivity, (Marie Curie coined this term for phenomenon). Professor Becquerel suggested that Marie Sklodowska Curie choose as the subject of her doctoral research project the identification of the unknown radioactive impurity in the uranium ore, pitchblende. Marie, with the help of her physicist husband Pierre, started with about 50 cubic feet of pitchblende ore, worked with batches as large as 40 pounds at a time, and stirred boiling mixtures in cast-iron basins with iron bars. By these heroic measures they succeeded in isolating two new elements from pitchblende that were more radioactive than uranium. The first they named *radium* for obvious reasons, the second *polonium,* after Marie's native Poland. Radium was 60 and polonium 400 times more radioactive than uranium. The yield was approximately one part of radium from ten million parts of ore. The Curies announced the discovery of radium and polonium in 1898, only two years after Becquerel's discovery of natural radioactivity.

Marie and Pierre Curie shared the Nobel Prize in Physics with Becquerel in 1903; Becquerel was granted half of the prize "for his discovery of spontaneous radioactivity" and the Curies, half "for their joint researches on the radiation phenomena discovered by Professor Henri Becquerel."

▼ POSTSCRIPT

Antoine Henri Becquerel was a famous offspring of a famous person. Both his father and his grandfather were distinguished scientists; both held the chair of physics at the Musée d'Histoire Naturelle in Paris. Born in 1852, Henri attended the École Polytechnique, after the conventional early training, and received the degree of Doctor of Science from that institution. He served as an engineer in the French government department of roads and bridges, but meanwhile taught physics at the museum where his father and grandfather had taught. Upon the death of his father in 1892, he succeeded to the chair his father and grandfather had occupied in the museum. In 1895 he was appointed professor of physics at the École Polytechnique. Only a year later he made the discovery for which he is most famous. He continued studies in the new and important field of radioactivity until his death in 1908.

In 1911, Marie Curie was awarded the Nobel Prize in Chemistry. Pierre had died in a traffic accident in 1906 or else he would have shared

▼ *Henri Becquerel, 1852–1908*

▼ *Marie Curie in her laboratory, 1912, the year after the award of her second Nobel Prize*

this prize also; Marie succeeded him as professor at the Sorbonne. Marie's citation read: "to Professor Marie Curie, Paris, for her services to the advancement of chemistry by the discovery of the elements radium and polonium, by the isolation of radium and the study of the nature and compounds of this remarkable element." Marie Curie died in 1934 of leukemia, the cancer undoubtedly a result of her exposure to the radiation whose danger was not fully appreciated until later.

Artificial Radioactivity and Nuclear Fission

Becquerel's discovery of natural radioactivity led to a new era, the atomic and nuclear age, but not right away.

Natural radioactivity was not understood for many years. It took the research and the insight of Rutherford and Soddy and others to deduce the nature and the source of the radiation that Becquerel and the Curies discovered to propose that it consisted of particles (α and β) emitted from the nuclei of atoms and the high-energy electromagnetic radiation associated with such emissions. Lord Rutherford was largely responsible for showing that the mass of atoms resides mainly (over 99.9%) at their centers, and is composed of positive and neutral particles called protons and neutrons. The secret of the means of unlocking the enormous power of nuclear energy began to be unfolded about 1934.

In 1934 Irene Curie, daughter of Marie and Pierre, and her husband, Frederic Joliot, discovered artificial radioactivity. They showed that the α-particles identified by Rutherford as parts of the atomic nuclei ejected by naturally radioactive elements could be used to bombard nonradioactive elements and induce these elements to be radioactive. Professor Alan Lightman in *Science 84* described this subatomic process: "Apparently, certain stable atomic nuclei, content to sit quietly forever, could be rendered unstable if they were obliged to swallow additional subatomic particles. The forcibly engorged atomic nuclei, in an agitated state, began spewing out little pieces of themselves, just as in 'natural' radioactivity."

Enrico Fermi, then in Rome, decided to use neutrons, rather than α-particles, to bombard stable elements. He attacked the nucleus of the massive uranium atom (the stable isotopic form of the element) in this way. He assumed that the neutron bombardment would probably create nuclei of elements close in weight to uranium. But Otto Hahn and Fritz Strassman, at the Kaiser Wilhelm Institute in Berlin, found among the products of the bombardment of uranium some barium, an element whose atoms are about half the size of those of uranium. Because there had been no barium in the sample bombarded, apparently some uranium nuclei had been cut in two!

In December 1938, Hahn wrote to Lise Meitner describing this unexpected result. Meitner had been a valued co-worker of Hahn for 30 years, but being Jewish, she had fled Hitler's Germany for Sweden five months earlier. At Christmas, her nephew Otto R. Frisch, also a physicist and a co-worker of the great Danish physicist Niels Bohr in Copenhagen, paid her a visit, and they discussed Hahn's letter. After much puzzlement as they walked in the snow, they remembered a theory of Bohr; in 1936 Bohr had suggested that the particles in an atomic nucleus might act in a collective way so that the nucleus might be deformed from a spherical shape by an impact from even a small particle like a neutron. The

▼ *Lise Meitner and Otto Hahn in their laboratory*

repulsive forces in the nucleus might overpower the attractive forces and the nucleus would split in two, sending the two halves flying apart at great speed and with a tremendous release of energy. He drew an analogy between an unstable heavy atomic nucleus and a rupturing water drop.

When Frisch went back to Copenhagen a few days later, he managed to speak to Bohr just as he was boarding the Swedish-American liner MS *Drottningholm* for New York. Bohr immediately sensed the importance of the nuclear *fission*—a term Frisch coined by analogy to cell division in biology—experiment observed by Hahn and of the Meitner/Frisch interpretation of it, including his own water-drop analogy. Bohr was on his way to a conference in Washington, D.C., on theoretical physics, and he carried to that conference the interpretation of Meitner and Frisch. Bohr subsequently wrote a short letter to the editor of *Physical Review* in which he outlined the liquid droplet theory of nuclear fission.

Nuclear fission was soon realized in chain reactions by Leo Szilard at Columbia. Bohr, who was now at Princeton, calculated that only a rare form of uranium, the isotope called U-235, present in nature to the extent of only about 1% of natural uranium, could sustain a chain reaction. To build a chain reactor required concentrating the U-235; it could be done, and was. (The use of Teflon in one of the two processes employed by the United States government to concentrate U-235 is described in the story on Teflon in Chapter 27.) Fortunately it was done in the United States before it could be done in Germany.

▼ POSTSCRIPT

On August 2, 1939, Albert Einstein sent a warning letter to President Franklin D. Roosevelt in which he said: "Some recent work by E. Fermi and L. Szilard . . . leads me to expect that the element uranium may be turned into a new and important source of energy in the immediate future . . . and it is conceivable . . . that extremely powerful bombs of a new type may thus be constructed."

The rest of the atomic energy program is well known: the manufacture and use of atom bombs in World War II; nuclear fusion, an even more powerful form of nuclear energy discovered and developed. Atomic energy plants for peaceful use have been constructed, but atomic energy has not yet performed up to expectations in the United States. Disposal of nuclear waste remains a serious problem, and coal and oil remain the principal sources of energy in the twentieth century.

What nuclear energy has changed dramatically and irrevocably is the meaning of war. Each new weapon throughout history seemed a great advance over its predecessors—the Roman catapult, the English longbow, the Swiss crossbow, gunpowder, nitroglycerin—but these strides were Lilliputian compared to the leap from even modern conventional weapons to nuclear bombs and missiles. The Federation of Atomic Scientists, an association of the scientists who worked on the atom bomb during World War II, later changed its name slightly, to the Federation of American Scientists, and now has as its main aim nuclear disarmament, convinced that nuclear weapons make war inconceivable. One is reminded of Alfred Nobel's response to Bertha von Suttner and her pacifist conferences: "My factories may make an end of war sooner than your congresses."

Let us hope that our generation and its handling of nuclear energy will be more successful than Nobel and his successors were in outlawing war.

22 ▼

SUBSTITUTE SUGAR: How Sweet It Is —and Non-Fattening

▼ The three most common substitutes for sugar were all discovered by accident. Saccharin, the first artificial sweetener, was discovered over a hundred years ago, long before it became fashionable to use a substitute for the common table sugar, sucrose. This happened in the laboratory of Ira Remsen, the most famous American chemist of the nineteenth century.

Remsen was born in New York in 1846; he went to Germany for graduate study at the universities of Munich, Göttingen, and Tübingen. Returning to the United States, he became professor of chemistry at Williams College, then at Johns Hopkins University. He established the first chemistry department in the United States of a quality equal to those of Europe and he counted among his students many future leading American chemists. He later became president of Johns Hopkins University. One of Remsen's students was my scientific "great-great-grandfather": E. P. Kohler. One of his students was James B. Conant, whose student was Louis F. Fieser, whose student was Charles C. Price, whose student was Royston M. Roberts. I sometimes like to point out that I can trace my chemical ancestry back to Wöhler, the father of organic chemistry (see Chapter 9) because Remsen was a student of Rudolph Fittig, who was a student of Friedrich Wöhler.

In 1879, one of Remsen's students was pursuing a problem assigned as part of an ongoing theoretical research program. While doing this, the student, who was named Fahlberg, noticed that a substance that he had prepared and accidentally spilled onto his hand tasted unusually sweet. (Chemists were not nearly so cautious then about smelling and tasting the materials they worked with as they are now.) Fahlberg apparently foresaw the possible importance of the new sweet-tasting substance, because he developed a commercial synthesis and took out a patent on it in 1885. The name he chose for it was saccharin, from the Latin word for sugar, *saccharum*.

In 1937 a chemistry graduate student working with Professor L. F. Audrieth at the University of Illinois was preparing a series of compounds called sulfamates because they were expected to have certain interesting pharmacological (not sweetening) properties. The student, Michael Sveda, noticed a markedly sweet taste to a cigaret he was smoking while in the laboratory and traced the source to one of the substances he was preparing, which turned out to be sodium cyclohexylsulfamate. The corresponding calcium salt was also found to be sweet. Both the sodium and calcium salts of cyclohexylsulfamic acid were used as sugar substitutes, the calcium salt being useful in low-sodium diets. These were widely used as sugar substitutes until 1970, when they were banned for use in the United States by the F.D.A. on the basis of animal tests. After the initial discovery of the sweetness of sodium cyclohexylsulfamate, a host of closely related sulfamates were prepared and tested, but none of them was as sweet as the first one recognized by the observant Sveda. Cyclamates were mixed with saccharin because the two together tasted sweeter and had a less bitter aftertaste.

The third important substitute sweetener, aspartame (sold as NutraSweet), was also discovered entirely by accident. The correct chemical name for aspartame is L-aspartyl-L-phenylalanine methyl ester. The "methyl ester" part of the name means that this substance is a close chemical relative of the dipeptide L-aspartyl-L-phenylalanine. A dipeptide is a combination of two of the amino acids that are the building blocks of proteins; when we digest a protein, it is broken down into the component amino acids. This methyl ester of the dipeptide was an intermediate that the Searle chemists had prepared in the process of making a tetrapeptide (a combination of four amino acids). The tetrapeptide was desired as a biological standard in connection with an anti-ulcer project.

One of the chemists accidentally tasted some of the intermediate dipeptide ester and discovered its remarkably sweet taste. The sweet taste of aspartame could not have been predicted from a knowledge of the properties of the component amino acids—one of them has a "flat" taste and the other is bitter. The extremely sweet taste that resulted from the combination of the two and conversion to the methyl ester was a complete surprise.

In *Aspartame: Physiology and Biochemistry* (1984), chemist James M. Schlatter describes the actual discovery of aspartame:

> In December 1965 I was working with Dr. Mazur on the synthesis of the C-terminal tetrapeptide of gastrin. We were making intermediates and trying to purify them. In particular, on an occasion in December 1965, I was recrystallizing aspartylphenylalanine methyl ester (aspartame) which had been prepared . . . and given to me by Dr. Mazur. I was heating the aspartame in a flask with methanol when the mixture bumped onto the outside of the flask. As a result, some of the powder got onto my fingers. At a slightly later stage, when licking my finger to pick up a piece of paper, I noticed a very strong, sweet taste. Initially, I thought that I must have still had some sugar on my hands from earlier in the day. However, I quickly realized this could not be so, since I had washed my hands in the meantime. I, therefore, traced the powder on my hands back to the container into which I had placed the crystallized aspartylphenylalanine methyl ester. I felt that this dipeptide ester was not likely to be toxic and I therefore tasted a little of it and found that it was the substance which I had previously tasted on my finger.

Unlike saccharin and cyclamate, which are excreted unchanged, aspartame is metabolized into its constituent natural amino acids, which are further metabolized by the usual body pathways. Because Schlatter knew this much about the metabolism of peptides, he was bold enough to taste the material that had splashed onto the outside of his flask.

There has been much controversy over the safety of saccharin and cyclamate. Saccharin has been used for over 80 years with no apparent ill effects, but questions were raised in the 1970s because bladder tumors were found in rats that had been fed massive amounts of saccharin. A proposed ban by the F.D.A. was postponed pending evaluation of additional animal testing. Although the ban on cyclamate was imposed in 1970 in the United States, this sugar substitute is still available in many other countries.

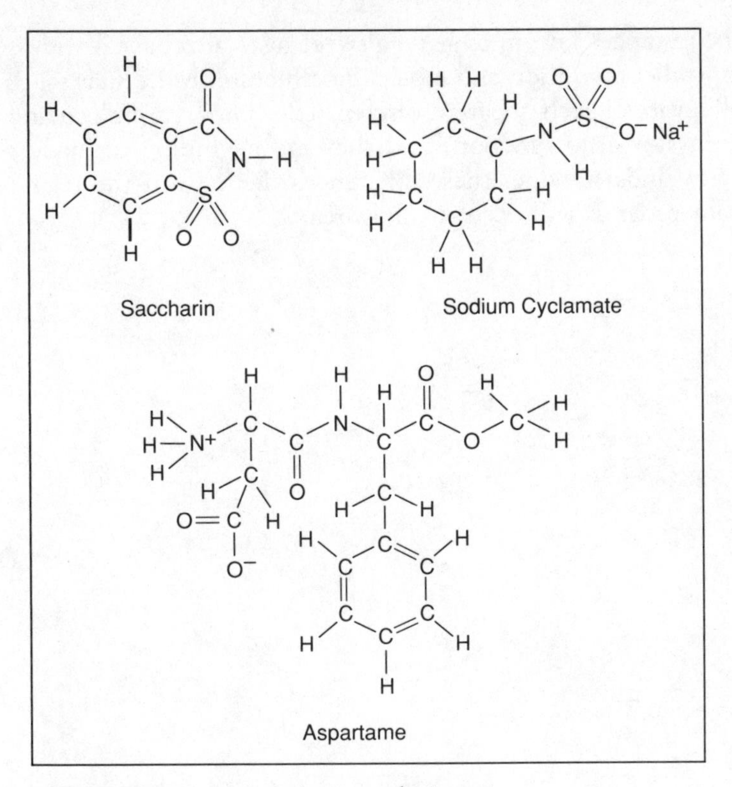

FIGURE 22.1. Formulas of sugar substitutes

It is difficult to give exact numbers to the degree of sweetness of a substance. Saccharin is admittedly the most powerful of all that are known. As normally used, it has a potency relative to sucrose of about 300 to 1. Cyclamate is about 30 times as sweet as sucrose and aspartame is about 200 times as sweet.

Other natural and synthetic nonnutritive sweetening agents have been used throughout the world besides the three described here, but these have been the most widely used in the United States. All three were discovered by accident, but perhaps these examples of serendipity are not as surprising as some others, because taste is notoriously subjective and unpredictable. Still, the molecular structures of saccharin, cyclamate, and aspartame are remarkably different, as shown in Figure 22.1. Although saccharin and cyclamate have a ring of six carbon atoms in common and both contain an atom of sulfur, aspartame is completely different—it has no elements of molecular structure in common with the other two.

The other substances known to have a sweet taste also have a wide variety of chemical composition and molecular structure. Molecular scientists are well aware of such apparent discrepancies between molecular structure and physiological function, and they are beginning to make some headway in understanding these differences. Perhaps in the near future significant progress will occur in this area.

23 ▼

Safety Glass

▼ The accidental discovery of safety glass occurred when it was most needed—soon after the invention of the automobile and the use of glass windshields. Automobiles were much more likely than horse-drawn buggies to go out of control and crash, causing serious injury to occupants from the broken windshields.

Natural glass, such as obsidian, has been present in nature since the forming of our planet. Obsidian and other forms of natural glass were formed from common elements in the earth's crust by intense volcanic heat followed by rapid cooling, long before anyone thought of changing their composition, color, or shape.

The origin of the first synthetic glass is lost in antiquity and legend. One of the most common legends was documented by Pliny the Elder, who lived in the first century A.D. A scholar and historian, he wrote 37 volumes of natural history. He died among the ashes from the eruption of Vesuvius in 79 A.D. after he had sailed with some of the Roman fleet under his command to the coast near Pompeii to help the inhabitants of that threatened area.

Pliny credited Phoenecian merchants with producing glass accidentally when they made a fire on a sandy beach. They were said to have rested their cooking pots on blocks of natron, a mineral of sodium carbonate probably brought from Egypt, and to have let the fire burn overnight to warm themselves. In the morning, to their amazement they found molten glass glistening among the ashes where the heat had fused the natron blocks with the silica of the sand. Although this accidental discovery and its date cannot be authenticated (it has been estimated at 4000 B.C.), the use of glass bottles by Egyptians as early as 1500 B.C. is well documented.

Glass has been around a long time. We know also that Romans used it for windows. In stationary positions in buildings it was generally used in many small panes with bronze frames, and even when used in small windows in carriages and buggies, it posed little danger in spite of its inherent fragility. With the advent of the horseless carriage, however, its use in windshields and windows became a potential source of injury.

In 1903 a French chemist named Edouard Benedictus dropped a glass flask on a hard floor. The flask shattered, but Benedictus noticed to his surprise that the fragments of glass did not fly apart, but the flask remained almost in its original shape. Benedictus examined the flask and found that it had a film on the inside to which the broken pieces of glass had adhered. He realized that this film had come from the evaporation of a solution of collodion (or cellulose nitrate, prepared from cotton and nitric acid), which the flask had contained. (Other serendipitous discoveries involving collodion have been described in Chapters 15 and 16.) On standing in the open glass flask, the solvent had evaporated, leaving a film of the collodion on the inside of the flask. Benedictus made a note of the incident on a label attached to the flask, but thought no more of it at that time.

After this laboratory accident, however, Benedictus read an account of a young girl who had been badly cut by glass in an automobile accident in Paris. Again, a few weeks later, he read of another such accident with serious consequences from flying glass, and it suddenly occurred to him that his experience with the nonshattering glass flask offered a potential solution to such problems. He rushed to his laboratory, found the labeled flask, and spent the night planning how a coating of some kind could be applied to make glass safe. It is said that by evening of the same day, with the help of a letter press, he had produced the first sheet of safety glass.

The name "triplex" coined for the new safety glass referred to the design of the material: it consisted of a sandwich in which two sheets of glass acted as the bread and the meat was a sheet of cellulose nitrate between them; the *three* sheets of transparent material were bonded together by heat. The development of the process from laboratory to factory production required several years and it was not until 1909 that Benedictus took out his first patent on the new safety glass.

Although Benedictus invented safety glass to prevent injury from flying glass from automobiles windshields, this was not the first practical use of the new laminated glass. It was first used in the lenses of gas masks in World War I. However, when the numbers of automobiles and their speed increased greatly during the 1920s, injury caused by glass became a concern, and laminated windshields became standard in American automobiles.

▼ POSTSCRIPT

Some readers might remember that the windshields in old automobiles tended to yellow with age. This is because the original plastic material used in the glass sandwich for safety glass was cellulose nitrate (collodion), which turned yellow from age and exposure to sunlight.

In 1933 the cellulose nitrate bonding material was replaced by cellulose acetate, which was more resistant to coloration by sunlight, but lacked strength over a broad temperature range and produced haze. Both cellulose nitrate and cellulose acetate were derived from cellulose, a material obtained from wood or other natural sources. Further investigation of possible plastic materials led to the finding that a completely synthetic polymer, poly(vinylbutyral) resin was better than cellulose acetate. Since 1939 this material has been the standard for laminated glass for automobiles, aircraft, and other applications requiring a strong transparent material.

Another form of safety glass is tempered glass that is not laminated—that is, it contains no plastic inner film. It shatters into many small pieces that are less likely to be damaging. It is used in side and back windows of automobiles, but in the United States and in some other countries, laminated glass is required for windshields.

Aircraft require great strength in windows because they must resist extremes of temperature and pressure and must also be resilient against high velocity bird impact. These requirements are met by highly spe-

▼ *The latest in safety glass technology: the non-lacerative shield*

cialized windshields that are complex, consisting of several layers of glass and plastic.

Recently, a second plastic layer has been used on the inside of the automobile windshield to prevent skin laceration by contact with the shattered glass that adheres to the center layer of plastic in the usual triplex laminate. In 1987 this was being offered on certain cars on a trial basis; preliminary results appear promising. In a head-on collision, a woman whose head hit a windshield that had the new antilacerative inner plastic coating suffered a severe bruise and a concussion, but no head or facial cuts.

In all of these types of safety glass except the tempered glass, the principle is the same as that observed serendipitously by Benedictus in 1903: containment of the glass particles by a film of plastic.

24 ▼

Antibiotics: Penicillin, Sulfa Drugs, and Magainins

Penicillin: Fleming, Florey, and Chain

▼ Perhaps the best-known important accidental discovery is Sir Alexander Fleming's discovery of penicillin. There is more serendipity in this discovery, however, than most people realize, and there are remarkable sequels to Fleming's discovery that ensured its importance, although these sequels are less well known.

Fleming's life is so full of apparently unrelated events, without any one of which it would not have reached the climax it did, that one "feels driven to deny their being due to mere chance," as his friend and colleague, Professor C. A. Pannett, said in his eulogy upon Fleming's death.

Alexander Fleming was born in rural Ayrshire, Scotland, in 1881. His father died when he was seven, leaving Alexander's mother to run the farm and raise four children of her own, plus some stepchildren. Alexander walked to a school a mile away when he was five, and when he was 10, he walked to a school four miles from home. When 12, the school was 16 miles away, so he boarded at Kilmarnock Academy, but walked a round trip of 12 miles each weekend to and from the train station to his home. After a year and a half at Kilmarnock, he went to London to join his older brother and resumed his schooling at the Polytechnic. This study was short-lived, however, because he could not afford it; the 16-year-old Fleming took a job with a shipping company, but still had time to join the London Scottish Volunteers. With this group he played on a water polo team, and at one time played against a team from St. Mary's Hospital, a part of the University of London.

A few years later he received a small legacy and his brother encouraged him to enter a medical school. There were 12 of these in London, and Fleming knew nothing about any of them—except the one affiliated with St. Mary's Hospital, which he knew had a water polo team, so there he went. At the same time Almoth Wright joined the school as a teacher in bacteriology. Fleming first planned to become a surgeon, but he was offered a position in (then) Sir Almoth Wright's laboratory following his graduation, and he worked in that laboratory the rest of his life, becoming Professor of Bacteriology in 1929.

During World War I Fleming and Wright were sent to France where they worked with wounded soldiers. Doctors at that time were depending on antiseptics to cure the battle wounds. But Fleming observed that phenol (or carbolic acid, the most common antiseptic at that time) did more harm than good, in that it killed the leucocytes (white blood cells) faster than it killed the bacteria, and he knew this was bad because the leucocytes are the body's natural defenders against bacteria.

In 1922 Fleming serendipitously discovered an antibiotic that killed bacteria but not white blood cells. While suffering from a cold, Fleming made a culture from some of his own nasal secretions. As he examined the culture dish, filled with yellow bacteria, a tear fell from his eye into the dish. The next day when he examined the culture, he found a clear space where the tear had fallen. His keen observation and inquisitiveness led him to the correct conclusion: the tear contained a substance that caused rapid destruction (*lysis*) of the bacteria, but was harmless to human tissue. The antibiotic enzyme in the tear he named *lysozyme.* It turned out to be of little practical importance, because the germs that lysozyme killed were relatively harmless, but this discovery was an essential prelude to that of penicillin, as we shall see.

In the summer of 1928, Fleming was engaged in research on influenza. While carrying out some routine laboratory work that involved microscopic examination of cultures of bacteria grown in petri dishes (flat glass dishes provided with covers), Fleming noticed in one dish an unusual clear area. Examination showed that the clear area surrounded a spot where a bit of mold had fallen into the dish, apparently while the dish was uncovered. Remembering his experience with lysozyme, Fleming concluded that the mold was producing something that was deadly to the staphylococcus bacteria in the culture dish. Fleming reported:

> But for the previous experience [with lysozyme], I would have thrown the plate away, as many bacteriologists must have done before. . . . It is also probable that some bacteriologists have noticed similar changes to those noticed [by me], . . . but in the absence of any interest in naturally occurring antibacterial substances, the cultures have simply been discarded. . . . Instead of casting out the contaminated culture with appropriate language, I made some investigations.

▼ ***Sir Alexander Fleming re-enacts examination of a petri dish "contaminated" by a penicillium mold.***

Fleming isolated the mold and identified it as belonging to the genus *penicillium,* and he named the antibiotic substance it produced *penicillin.* Later he would say, "There are thousands of different moulds and there are thousands of different bacteria, and that chance put the mould in the right spot at the right time was like winning the Irish sweep." The comment about the "thousands of different bacteria" is pertinent, because although penicillin is deadly to many bacteria, including staphylococcus, it has no effect on some other types of bacteria. Fortunately, the bacteria that penicillin kills are some of those responsible for many common and serious human infections.

The use of molds against infections was not totally novel in 1928. Louis Pasteur and his co-worker J. F. Joubert showed in 1877 that one microbe could prevent the growth of another. There are records of molds from bread being used by the Egyptians and the Romans in ancient times, but there are thousands of different molds that will grow on bread and only a few of them will produce anything useful against infection. Fleming must have known of this, and thus we can understand his amazement.

Fleming went on to show that penicillin was not toxic to animals and was harmless to body cells.

> It was this nontoxicity to leucocytes that convinced me that some day it would come into its own as a therapeutic agent. . . . The crude penicillin would completely inhibit the growth of staphylococci in a dilution of up to 1 in 1000 when tested in human blood, but it had no more toxic effect on the leucocytes than the original culture medium . . . I also injected it into animals and it had apparently no toxicity. . . . A few tentative trials [on hospital patients] gave favourable results but nothing miraculous and I was convinced that . . . it would have to be concentrated . . . We tried to concentrate penicillin but we discovered . . . that penicillin is easily destroyed . . . and our relatively simple procedures were unavailing.

Meanwhile the remarkable success of sulfanilamide had brought chemotherapy into prominence (see the following story on the sulfa drugs). Attempts by Harold Raistrick, in collaboration with Fleming, to isolate and concentrate penicillin were unsuccessful and nothing more was done about penicillin for several years. In the late thirties, Howard W. Florey, a professor of pathology at Oxford University, began a research collaboration with Ernst Boris Chain, a Jewish refugee biochemist from Hitler's Germany who had been brought to Oxford by Florey. They initiated research on lysozyme, the antibacterial enzyme discovered by Fleming, and other natural antibacterial substances. Their work quickly centered on penicillin as the most promising of these agents.

Using sophisticated chemical techniques of isolation and concentration that were available at Oxford and were familiar to Florey and Chain, but not to Fleming at St. Mary's, the Oxford team succeeded in concentrating and purifying penicillin to such an extent that they could demonstrate its curative properties, first on experimental infections in mice, and then on human patients suffering from staphylococcal and other serious infections. (The first penicillin used on humans was grown in hospital bed pans; some clinical tests were terminated prematurely owing to the scarcity of the drug, even though it was recovered from the urine of patients and reused.)

Because of the urgency of potential use against disease and the wounds of military personnel in World War II, production on a large scale became a prime concern, both in Britain and in the United States. Florey came to the United States to describe the methods of extraction and production used in Britain, and chemists on both sides of the Atlantic worked feverishly to determine the chemical structure of penicillin and to produce it by synthesis or by fermentation. This sensitive and complicated molecule was first synthesized long after the war, but progress in developing production by fermentation during the war was phenomenally rapid.

Serendipity entered into this phase of the production of penicillin, as well as in its discovery. When Florey came to the United States to consult on means of producing penicillin on a large scale, he visited the Northern Regional Laboratory of the U.S. Department of Agriculture in Peoria, Illinois. This laboratory had for some time been seeking an industrial use for surplus cereal crops and the solution to the related problem of disposing of a viscous extract that was a by-product of the corn milling process. When this extract was incorporated into the culture medium for penicillin, it unexpectedly increased the yield of the desired mold by a factor of 10.

A second contribution by the Peoria laboratory came from the development of an improved strain of penicillin-producing mold. Hundreds of molds from all over the world were collected and brought to Peoria for testing. Unbelievably, the winning contribution was made by a local woman named Mary Hunt, dubbed "Moldy Mary" because of her enthusiasm for searching for new mold sources. She brought a cantaloupe from a Peoria fruit market that had a mold with a "pretty, golden look." This new strain of mold doubled the yield of penicillin, so that the combination of the two discoveries at Peoria increased the yield of penicillin by 20-fold. Who could have predicted that Peoria would contribute so significantly to the production of the miracle drug discovered accidentally in London?

Not only were thousands of lives saved by penicillin during the war, but also research was stimulated for the discovery of other antibiotics, including a family of compounds related chemically to penicillin known as cephalosporins. Some of these newer antibiotics are effective against bacteria that are resistant to penicillin. (An account of the accidental discovery of cephalosporin C is given in Chapter 30.)

Fleming, Florey, and Chain shared the Nobel Prize in Physiology or Medicine in 1945. All three were subsequently knighted for their work, which had resulted in the relief of much suffering and the saving of uncounted lives.

Sir Alexander Fleming was aware of his encounters with serendipity. He once said, "The story of penicillin has a certain romance in it and helps to illustrate the amount of chance, or fortune, of fate, or destiny, call it what you will, in anybody's career." I hasten to add that if it had not been for Fleming's intelligence, or sagacity—to use the term that was an essential component of Walpole's definition of serendipity—the accidents that happened to Fleming would have come to nothing.

Sulfa Drugs: Domagk, Fourneau, and the Trefouels

The discovery of sulfanilamide and closely related synthetic bacteriocidal drugs came as much by way of misconception (see Chapter 35) as by serendipity, but it is a fascinating story. Gerhard Domagk was awarded the Nobel Prize in Physiology or Medicine in 1939 for the discovery, but many others played important parts in the drama.

Domagk was born in Lagow, Germany, in 1895. He entered the University of Kiel, but World War I interrupted his medical studies. After the Armistice he reentered the University of Kiel and was granted his medical degree in 1921. After some years as a lecturer at the University of Greifswald and as a professor at the Pathological Institute at Muenster, in 1932 he held a position with the I. G. Farbenindustrie, the German dye cartel, as Director of their Laboratory for Experimental Pathology and Bacteriology. Here his job was to test the pharmacological properties of the new dyes synthesized by chemists Fritz Mietzsch and Joseph Klarer. Mietzsch had distinguished himself by preparing the first successful synthetic antimalarial drug, Atabrine. Malaria is a protozoal infection, as is syphilis, for which Ehrlich had found salvarsan to be an effective drug—the first magic bullet to be used against disease by chemotherapy. But no chemicals were effective against bacteria in 1932—bacteria that caused the terrible diseases of pneumonia, meningitis, gonorrhea, and streptococcic and staphylococcic infections.

The I.G.F. team of chemists and pharmacologists set out to find compounds that would kill these microbes without harming their animal or human hosts. The plan they conceived was to prepare certain dyes, to see whether they might be bacteriocidal—because certain types of dyes, specifically those containing a sulfonamide group, seemed to be particularly "fast" (tightly bound) to wool fabrics, indicating an affinity for protein molecules. The chemists thought that because bacteria are protein these dyes might fasten themselves to the bacteria to inhibit or kill them. As we shall see, this theory was only partly correct: the sulfonamide group was essential, but the part of the molecule that made it a dye was nonessential to bacteriocidal effectiveness.

One of the dyes that Mietzsch and Klarer made and Domagk tested on laboratory mice and rabbits infected with streptococci was called Prontosil. It was found to be strongly disinfective against these bacteria and could be tolerated by the animals in large doses with no ill effects. This discovery of the bacteriocidal effect in animals was probably made in early 1932; I.G.F. applied for a patent in December of that year. Clinical tests (on human patients) apparently began soon after this, but the record is confused. Some accounts say that before any other tests had been made on humans, Domagk in desperation gave a dose of Prontosil to his deathly ill young daughter, who had developed a serious streptococcal infection following a needle prick, and the girl then made a rapid recovery. Others report that the first clinical test was on a 10-month-old baby boy who was dying of staphylococcal septicemia and was given Prontosil by his doctor, Richard Foerster. Foerster was an associate of Dr. Hans Schreus, a professor in the medical school in Duesseldorf and a friend of Heinrich Hoerlein, who was Domagk's superior at I.G.F. Consulted by Foerster about the sick baby boy, Schreus remembered that Hoerlein had told him about a red dye (Prontosil) that was miraculously effective in animals against streptococci. Feeling that there was nothing to lose if the dye was not effective against staphylococci, because the baby was close to death anyway, Foerster gave the child two doses of the red dye, and he made a rapid recovery from what was thought to be a fatal blood poisoning. Regardless of which of the two stories is correct, or whether both are true, it is certain that by the middle 1930s a medical miracle was widely recognized, on the basis of which Domagk was awarded the Nobel Prize in 1939.

There were, however, other important developments in the years between 1933 and 1939. Domagk did not publish the results of his tests of Prontosil on animal infections until February 1935, more than two years after the work was done, and a month after the patent on Prontosil was issued to I.G.F. A husband and wife team at the Pasteur Institute in Paris, J. and J. Trefouel, under the direction of M. Fourneau, learned of the German work and made a fundamental discovery. They examined several compounds closely related in chemical structure to Prontosil—"azo" dyes. The term *azo* for this particular kind of dye comes from the French word for nitrogen, *azote,* and is used because of the two nitrogen atoms joined by a double bond in the center of the molecule. These dyes differed significantly in one part of the molecule, but all contained the same sulfonamide part. The Trefouels found that these dyes have antibacterial properties almost identical with those of Prontosil.

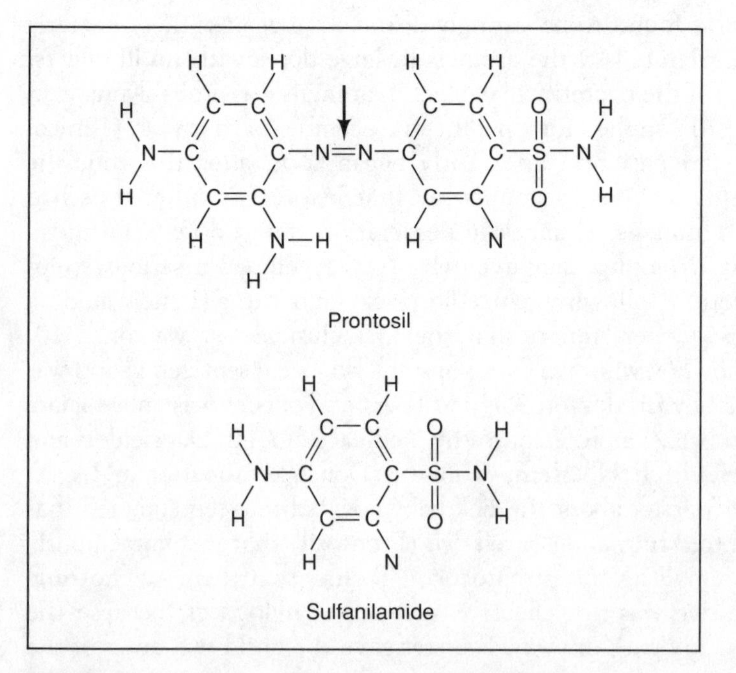

FIGURE 24.1. Formulas of Prontosil and sulfanilamide

This finding also explained another puzzle about Prontosil: it was ineffective *in vitro* although strongly effective *in vivo* against bacteria. Some function of the body made the sulfonamide dyes antibacterial. The French scientists reasoned that in the animal body the dye is broken into two parts, and only the sulfonamide half is the effective antibacterial agent. To prove this, they synthesized the simpler sulfonamide part of Prontosil, which was the known compound *p*-aminobenzenesulfonamide, called more simply "sulfanilamide", and they found it was as effective as Prontosil against bacterial infections. The formulas of Prontosil and sulfanilamide are shown in Figure 24.1. If the molecule of Prontosil is cleaved at the double bond between the middle nitrogen atoms (the position indicated by the arrow), and two hydrogen atoms are added to the right-hand part, sulfanilamide results. This cleavage occurs in the animal body when Prontosil is injected or imbibed, and the sulfanilamide so produced is the actual antibacterial agent. Sulfanilamide is colorless, as is the other half of the Prontosil molecule. Linking the two parts through the double bond between the two central nitrogen atoms gives the azo dye its color. The theory that sulfonamido dyes would be bacteriocidal was a misconception because only the sulfonamide part of the dye molecule is the microbe killer; the fact that it was a part of a dye molecule was incidental, and in this sense the discovery was serendipitous.

The Trefouels' discovery made the patent on Prontosil by I.G.F. useless. Although sulfanilamide had been synthesized and patented many years before as a dye intermediate, the patent had expired by the time the substance was found to be a potent bacteriocide.

Following the announcement of the findings by Fourneau from the Pasteur Institute in 1935, clinical trials of sulfanilamide were made in France, England, and the United States with miraculous success. A case that gave great publicity to the new drug was the use of Prontosil to save the life of Franklin D. Roosevelt, Jr., son of the President. In 1936 young Roosevelt was dying from a streptococcic infection, when, at the request of his mother, Eleanor Roosevelt, Dr. George Tobey at Massachusetts General Hospital gave him Prontosil. He then made a rapid recovery.

Many simple analogs of sulfanilamide were synthesized by chemists and tested on animals and then on humans; in fact, by 1947, over 5,000 sulfonamides related to sulfanilamide had been prepared and tested. Not all of these were effective, but some were found to be better than sulfanilamide against certain diseases. Sulfapyridine, for example, was prepared in 1938 and was found to be more effective against pneumonia, and sulfathiazole was produced and used medically by 1940.

▼ POSTSCRIPT

The molecular structure of the effective sulfonamides is remarkably similar. Of the thousands of compounds prepared and tested, almost all of the active ones are those in which the only variation in structure is a change in the group of atoms attached to the right of the nitrogen atom in boldface in the formulas of sulfapyridine and sulfathiazole shown in Figure 24.2.

The sulfa drugs were used with great effectiveness in the 1940s, especially by the armed forces during World War II. They were largely superseded by penicillin and other modern antibiotics, but they are still useful in the treatment of some diseases. One of their drawbacks was their insolubility, which led to deposition in the kidneys with resulting renal damage. It was found that this difficulty could be alleviated by administration of a combination of three different sulfa drugs, the combined dose being as effective as an equal amount of one of them, but the concentration of each being one third, so that they were excreted satisfactorily.

I mentioned earlier that Domagk was awarded the Nobel Prize in 1939. This is not quite correct. Although he was selected for the prize by the authorities in Stockholm, he did not actually receive it until many

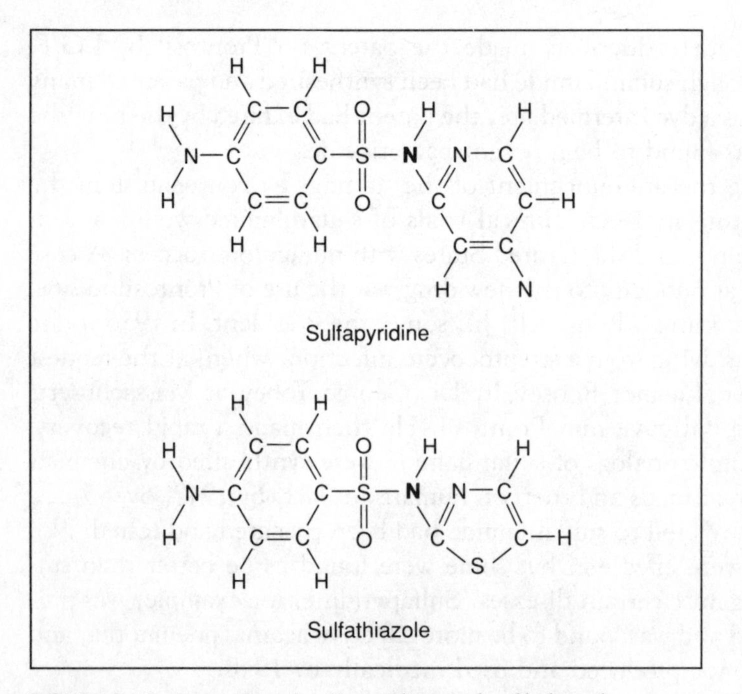

FIGURE 24.2. Formulas of sulfapyridine and sulfathiazole

years later. When he received notice of the award in October of 1939, Domagk sent a letter of acceptance, but a second letter declining it was received in Stockholm in November. The second letter was the result of pressure from the Nazis; Domagk was by this time in the hands of the Gestapo. After the war, in 1947, Domagk was able to visit Stockholm, deliver the Nobel lecture, and receive the medal and diploma—but not the prize money, which had reverted to the Nobel Foundation funds.

Magainins: Antibiotic from a Frog's Skin

Dr. Michael Zasloff, a pediatrician and biochemist at the National Institute of Child Health and Human Development, stared at the frog in the tank in his laboratory. It was the summer of 1986. He had operated on the frog a few days earlier to remove its ovaries as a part of his study of lung infections of children born with cystic fibrosis. What suddenly amazed him was that the frog's surgical wound was clean, closed, and healing rapidly in the murky water of the laboratory aquarium, which teemed with bacteria that should have caused an infection. This was not the first such occasion for Dr. Zasloff. Like many other scientists, he had operated on this kind of African frog many times, and the frogs' wounds had healed well each time. But it was the first time that Dr. Zasloff realized that he was observing a miracle, and he wondered how it happened.

In subsequent research, Dr. Zasloff isolated two peptides (fragments of proteins) from frog skin and determined the sequence of amino acids in each. These peptides were highly active against a variety of bacteria, fungi, and protozoa, such as the malarial parasite. Dr. Zasloff named them *magainins,* after the Hebrew word for shield. Dr. Zasloff and his colleagues have isolated the magainin genes from frogs and will use them to probe for similar genes in humans.

Although this discovery was not in Dr. Zasloff's plans at all, his mind was ripe for it. He had long been preoccupied with two related subjects. One was curiosity about why children born with cystic fibrosis, a hereditary disease, develop frequent severe lung infections from bacteria not found in the lungs of healthy individuals. The other was his fascination with the implications of the discovery several years ago of cecropins, substances found in some insects that appear to confer powerful, natural protection against bacteria. Cecropins are peptides, and they appear to kill bacteria by disrupting the membranes of bacterial cells without harming the membranes of the insects' own cells. The magainins are the first chemical defense system separate from the immune system to be found in vertebrate animals.

25 ▼

NYLON: Cold Drawing Does the Trick

▼ There is a saying that "it is better to be lucky than smart." According to an old Norse saga, the Vikings were willing to risk their lives following Leif Ericson across the wild North Atlantic in open boats not because he was strong, courageous, or smart (although he may have been all of these), but because they thought he was lucky. In fact Leif means lucky; "Leif Ericson" is the old Norse way of naming him Lucky son of Eric. He was the son of Eric the Red, the founder of the earliest Scandinavian settlements in Greenland. Leif Ericson's discovery of North America ca. 1000 A.D. is another example of serendipity! Leif had been commissioned by the King of Norway to proclaim Christianity in Greenland but he was driven far off his course from Iceland to Greenland by bad weather and landed in Newfoundland instead.

The story of nylon seems to bear out this saying about luck. Wallace Hume Carothers was brought to Du Pont to direct their new basic chemical research program because his colleagues at the University of Illinois and Harvard University recommended him as the most brilliant organic

▼ *Wallace H. Carothers in his laboratory at the Du Pont Company's Experimental Station near Wilmington, Delaware*

chemist they knew. Carothers initiated a program aimed at understanding the composition of natural polymers such as cellulose, silk, and rubber, and to produce synthetic materials like them. Although by 1934 his group had contributed valuable fundamental knowledge in these areas, he had just about decided that their efforts to produce a silk-like synthetic fiber had failed, when an accident occurred during some horseplay among his chemists in the laboratory. This accident turned the failure into the enormous success advertised at the 1939 New York World's Fair as "Nylon, the Synthetic Silk Made from Coal, Air, and Water!"

In an article in the July 1981 *Journal of Chemical Education,* Dr. Carl S. ("Speed") Marvel described the accident that led to the commercial development of nylon. (There are several stories about how Speed Marvel got his nickname. When I was a graduate student at the University of Illinois, I heard that he had completed in six weeks a laboratory course in organic analysis that was expected to take a full semester. Speed himself, however, demurred from this explanation and claimed the name came from the fact that he could get up later than anyone else in the chemistry fraternity house and still beat everyone to the breakfast table!)

> Nylon (a polyamide, having a structure similar to that of silk) had been made and seemed not to have any especially useful properties and [it was] put aside on the shelf without patenting. Work was continued on the polyester series, which gave more soluble products, easier to handle, and

▼ ***Julian Hill, colleague of Carothers, re-enacts the discovery of the cold-drawing effect for the production of strong fibers of nylon***

> thus simpler to work with in the laboratory. It was while working with one of these softer materials that Julian Hill noted that if he gathered a small ball of such a polymer on the end of a (glass) stirring rod and drew it out of the mass, that it was extended and became very silky in appearance. This attracted his attention and that of the others working with him, and it is reported that one day while Carothers was downtown, Hill and his cohorts tried to see how far they could stretch one of these samples, and [they] took a little ball on a stirring rod and ran down the hall and stretched them out into a string. It was in doing this that they noticed the very silky appearance of the extended (strands) and they realized that they were orienting the polymer molecules and increasing the strength of the product.

Because the polyesters they were working with had melting points too low for use in textile products, they went back to the polyamides that they had put aside and found that these fibers, too, could be "cold drawn" to increase their tensile strength so much that they made excellent textiles. Filaments, gears, and other molded objects could also be made from the strong polymer produced by cold drawing. Dr. Marvel told me that Du Pont never had a "composition of matter" patent on nylon, but only patented the cold-drawing process. This process, discovered accidentally, led to the most important product Du Pont ever put on the market. The horseplay incident led to a discovery of worldwide economic and social importance, however, only because of the insight that Julian Hill and his associates displayed in recognizing the scientific implications of the appearance and physical properties of the stretched strands of polyester they played with.

Figure 25.1 shows how the cold-drawing process works. The most common nylon is called nylon-6,6. The sixes refer to the number of carbon atoms in the two monomer units, one of which is a six-carbon diacid (containing oxygen atoms) and the other a six-carbon diamine (containing nitrogen atoms). When the two monomers combine chemically to produce the polymer, they do so by eliminating a molecule of water (H_2O) between each end of each monomer unit, and the two monomer units then alternate in the chain, as you can see in the top part of the figure. In the cold-drawing process, the long polymer molecules are lined up with one another in such a way that every oxygen atom on one polymer chain can form a hydrogen bond with a nitrogen atom on an adjacent chain. This binds the individual polymer molecules together in much the same way separate strands in a rope, when twisted together, form a cable, and this association of linear polymer molecules through hydrogen bonding is responsible for the greatly increased strength of the

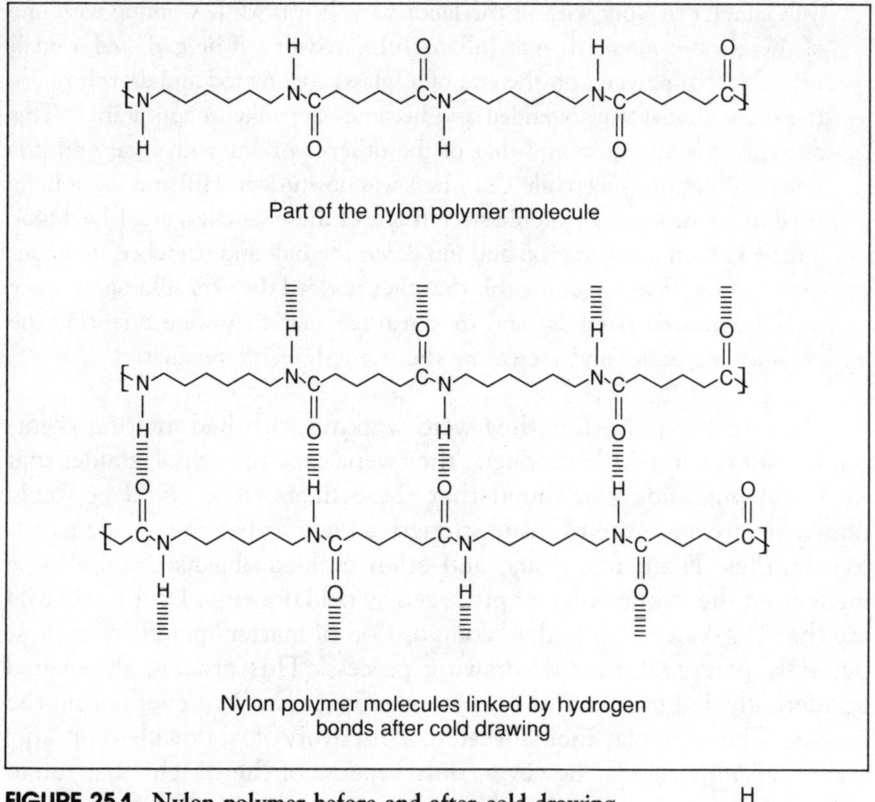

Part of the nylon polymer molecule

Nylon polymer molecules linked by hydrogen bonds after cold drawing

FIGURE 25.1. Nylon polymer before and after cold-drawing. Every corner in the zigzag line represents a carbon atom with two hydrogen atoms attached. Only the atoms involved in the intermolecular linkages are shown by their symbols.

nylon fibers. We think that the same principle accounts for the strengths of silk fibers; the natural polyamide molecules of silk are oriented in such a way that hydrogen bonds hold the individual molecules together. Silk worms accomplish the cold drawing as they extrude the viscous silk filaments.

▼ POSTSCRIPT

Although the cold-drawing process that made nylon a successful synthetic fiber was discovered with polyesters, Carothers's polyesters were poor textile fibers. Shortly after the introduction of nylon as a synthetic fabric, British chemists discovered a polyester that *was* strong enough to make an excellent fiber, and they put it on the market under the trade name Terylene. Although Carothers's chemists made many polyesters from diacids and dialcohols, one diacid they did not try was terephthalic acid—the one diacid that produces good fibers!

Du Pont still managed to profit from the polyester work, however, because their film department found that the polyester the British used in Terylene could be used to make a strong film if it was stretched biaxially (presumably the same principle was involved as in cold drawing a filament). The film was marketed as Mylar, and Du Pont produced their version of Terylene as Dacron at about the same time. On a visit to Du Pont many years ago I heard an interesting story about the naming of their polyester fiber. They held a company contest to choose the name. The night before the name was to be announced and the prize presented, the person in charge of the contest became anxious and decided to check again to make certain that the chosen name was not already copyrighted. Sure enough, the final check revealed that it *was.* In the confusion that followed, the second-place name Dacron, was chosen, without a thorough check. Although not already used for something else, it violated a cardinal Du Pont rule: A trade name should not be capable of being mispronounced. Consider Nylon and Orlon, for example. It is hard to think of more than one way to pronounce these words. But Dacron can be pronounced with either a long or short *a.* I was told this story when I asked someone at Du Pont, "What *is* the correct pronunciation of Dacron?"

Dacron became a popular component of textile fabrics for men's clothing, alone and in blends with wool. Mylar film has been used since the 1950s for microfilm and magnetic audio film, and recently has become the preferred material for compact disks. A recent interesting application has been as a very thin, light, and strong fabric covering the wings of Daedalus, the human-powered aircraft that set a new distance record, flying from one island to another across the Aegean Sea in April 1988.

The introduction of nylon at the 1939 New York World's Fair was one of the most spectacular consumer events of all time. The Du Pont exhibit graphically explained the chemical synthesis of nylon from "coal, air, and water" and nearby was an attractive model in a gigantic glass "test tube" displaying the new hose very visibly. I was a young chemist and the display certainly captured my attention!

When "nylons" were first offered for sale in New York City on May 15, 1940, four million pairs of hose were sold in the first few hours. Such sales were not to be repeated for very long, however, for war clouds were

▼ ***Movie star Betty Grable auctions her nylon hose for $40,000 at a war bond rally. A promotion encouraged women to donate nylon hose to be melted to make parachute fabric.***

already gathering, and no sooner had the American woman become enamored of the new fabric than it was taken away for war applications, mainly parachute material. Not only were there no new "nylons" to be purchased, but women who had them were encouraged to turn them in to be converted into parachutes.

26

Polyethylene: Thanks to Leaky and Dirty Equipment

Teflon (polytetrafluoroethylene) was discovered just before World War II and developed because of its usefulness in the production of atomic bombs. Another polymer, also discovered in the 1930s, was enormously important to the Allies in the war effort; this was polyethylene, which was used for insulation of radar cables and was largely responsible for the effectiveness of this new electronic instrument. I mention in the Friedel–Crafts story that one practical application of the discovery coming from the French-American collaboration was the production of the high-octane aviation fuel that gave the British fliers superiority over the German aircraft. Consideration of Teflon, polyethylene, and aviation gasoline—just these three applications of chemical discoveries and developments—can well justify the pride of Allied chemists in their contributions to winning World War II.

The discovery of polyethylene, which the British called polythene, is usually credited to the British chemists at Imperial Chemicals Industries (I.C.I.), where it was produced accidentally and developed soon afterward for an important wartime application. The serendipitous discovery is described by J. C. Swallow, a chemist at I.C.I. in Chapter 1 of *The History of Polythene* (1960):

> As time passes any story tends to become idealized and presented in a way which suggests a steady and logical growth from the start of the research programme to the discovery and development of the particular product. The story of polythene, however, provides an unusually clear-cut instance of the unexpected results that may come from research, and of the importance of the role of chance in such work. . . . The actual history of polythene began in 1932 at Northwich, Cheshire, in the Alkali Division of I.C.I., when M. W. Perrin and the author of this chapter [Swallow] recommended that work on the effect of very high pressures on chemical reactions should be carried out. . . . Some fifty reactions were tried during 1932 and 1933 and the results were all disappointing; [i.e., none gave any interesting or valuable products]. Amongst [them] was the reaction between ethylene and benzaldehyde, which was tried in March 1933, at 170°C and with an ethylene pressure of 1,400 atmospheres [a very high pressure]. At the end of the experiment, the walls of the [reaction] vessel were found to be coated with a thin layer of a "white waxy solid" to quote from the notebook of R. O. Gibson, who carried out the experiment. The solid was recognized as a polymer of ethylene, but on repeating the experiment with ethylene alone the ethylene decomposed with great violence.

Swallow reports that they discontinued work at high pressures until they could design and build better equipment. In December 1935 further experiments with ethylene were carried out using improved apparatus. When a temperature of 180°C was reached, the pressure dropped, so more ethylene was pumped in. When they opened the small reaction vessel, they found a total weight of eight grams of white powdery solid. They recognized that the polymerization of the gaseous ethylene to the solid polymeric form could not have accounted for all of the pressure drop observed, and they suspected a leak in one of the joints of the apparatus.

Swallow writes:

> Here again the element of chance played an important role, and it took some months of intensive work by all those in the research team to elucidate the full reasons as to why, if the leak had not occurred, the experiment would probably have been far less spectacular than it was, and might have been a repetition of the earlier [unsuccessful] ones.
>
> The success of the experiment in December was in fact due to the additions to the reaction vessel of fresh ethylene to replace that which had

leaked out. This ethylene contained by chance about the right amount of oxygen to catalyse the formation of successive amounts of the polymer.

Swallow also writes that they found that polyethylene exhibited the phenomenon of "cold drawing" observed by W. H. Carothers of Du Pont as a property of polyesters and polyamides (described in Chapter 25) which indicated that this polymer of ethylene was a reasonably straight-chain polymer of fairly high molecular weight. However, the I.C.I. people saw no good use for the new polymer, and it would probably have remained a laboratory curiosity but for the role that chance played again.

J. N. Dean of the British Telegraph Construction and Maintenance Company, which was associated with the production of underwater telegraph and telephone cable, heard about the new polymer and recognized that it might be a good insulator of underwater cables. (There is an almost uncanny parallel in the way in which polyethylene came to be developed for radar by the British through coincidental communication between Mr. Dean and I.C.I. and how Teflon came to be developed for the atomic bomb work by the United States through the coincidental communication between General Groves and Du Pont. See the Teflon story in Chapter 27.) He expressed great interest in examining the material even though it was available in only small quantities. In addition, another I.C.I. employee had recognized the similarity of the mechanical properties of polyethylene to those of gutta percha (a natural polymeric material related to rubber, but nonelastic), which had been used for insulation of telegraph cables.

In July 1939 I.C.I. made enough material for one nautical mile of underwater cable, and it was found satisfactory for this purpose, although the outbreak of war prevented full trials. The experience gained in producing this cable turned out to be invaluable for the quick developments needed a few years later. On the basis of expected needs for submarine telephone cable, a small plant was built that came into operation in September 1939 on the day the Germans invaded Poland. Actually, underwater cables insulated with polyethylene were laid near the end of the war, when they formed a useful link between England and France.

However, at the beginning of the war, there arose an immediate need for flexible high-frequency insulated cable for ground and airborne radar equipment. Polyethylene for this purpose was produced by I.C.I. in England and by Du Pont and Union Carbide in the United States.

The significance of polyethylene in radar equipment is illustrated dramatically by a statement (quoted in *Polythene* by Swallow) of Sir Robert Watson Watt, the discoverer of radar, who said in August 1946:

> The availability of polythene transformed the design, production, installation, and maintenance problems of airborne radar from the almost insoluble to the comfortably manageable. [—owing to its almost ideal properties of insulation and structural integrity] . . . . A whole range of aerial and feeder designs otherwise unattainable was made possible, a whole crop of intolerable air maintenance problems was removed. And so polythene played an indispensable part in the long series of victories in the air, on the sea, and on land, which were made possible by radar.
>
> Polythene was an essential element in that "single technical device" to which the Fuehrer ascribed the "temporary" (but as it proved, enduring) set-back experienced by his U-boats.
>
> It made its contribution to the major naval combats typified by the action in which, as the Commander-in-Chief said, radar enabled the Home Fleet to "*find, fix, fight, and finish the Scharnhorst.*" It had its part in such continuing operations of the smaller naval crafts.
>
> It had its vital place in the small batch of sets of anti-U-boat airborne radar equipment which, with their shipborne counterparts—also polythene aided—permitted the sinking of a hundred U-boats within a very few weeks.
>
> And centimetric aerial systems in polythene moulding multiplied the effectiveness of our bomber force by a very large factor indeed. . .

Polyethylene, like Teflon, was little known by the general public until after the war. Then the large-scale developments were in the manufacture of film and of molded objects. Film production developed rapidly in the United States where polyethylene began to replace cellophane in many applications. Films of polyethylene are now used for packaging produce, frozen and perishable foods, textile products, and merchandise of all kinds. Other film uses include extensive applications in agriculture (greenhouses, ground cover, tank, pond, and canal liners), construction (moisture barriers, utility covering materials, and drop cloths), to say nothing of the indispensable garbage bag. Insulation of electrical cables has continued to be an important use of polyethylene. It was the first plastic to exceed one billion pounds of production per year (in 1959).

The kind of polyethylene used in common applications, produced under pressure with catalysis by oxygen essentially as discovered by the I.C.I. chemists, is low-density, branched polyethylene. "Branched" refers to the nature of the polymeric chain made up of two-carbon ethylene monomer units. If the ethylene molecules became linked to one another completely end to end, the polymer molecule would be *unbranched,* as shown in Figure 26.1. However, the polymer produced until the middle 1950s had some branches, like a tree limb. The branching made it lower-melting and less dense (lighter) than the high-density, linear (unbranched) polyethylene that was ultimately produced in the 1950s by new methods.

The accidental production of polyethylene by the I.C.I. chemists in 1933 was not the first time this happened. About three years earlier the polymer also came uninvited and unexpected as a by-product in an experiment carried out by M. E. P. Freidrich while a graduate student under the direction of Professor Carl S. Marvel at the University of Illinois. The experiment involved the use of ethylene in the presence of a lithium alkyl compound at ordinary atmospheric pressure.

In a video tape made in 1980 by the American Chemical Society Division of Chemical Education, Professor Marvel described the 1930 incident as his "introduction to polymer research," to which he made major contributions for over 50 years. He commented that he did not follow up on this observation immediately, however, because "nobody thought polyethylene was good for anything."

This reminds me of another classic example of misjudgment or, perhaps, inability to predict the future. Many years ago I happened to talk to Professor Emmett Reid, of Johns Hopkins University, who told me he was the first to discover how to prepare ethylbenzene in good yield from ethylene and benzene by a Friedel–Crafts reaction. (See Chapter 17.) Reid did not bother to patent the process, because, he said, "ethylbenzene wasn't worth anything." This was indeed true in 1928. How-

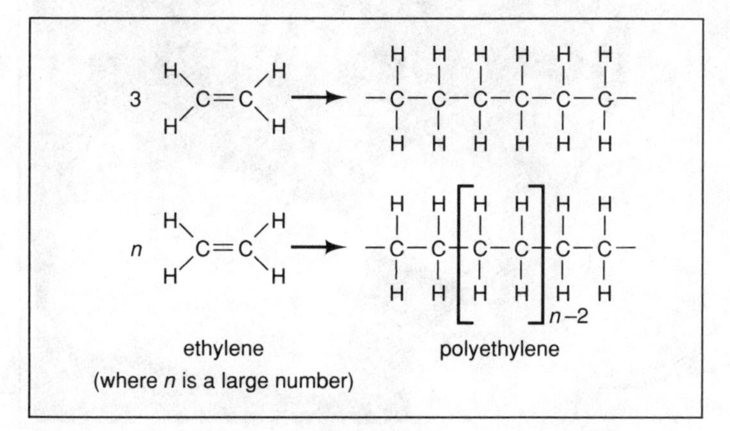

Figure 26.1. Linkages of ethylene molecules to form unbranched polyethylene

ever, ethylbenzene is the precursor to styrene, and in the 1930s it was found that styrene could be converted into polystyrene. Polystyrene is a very useful and versatile material; in 1987 over five billion pounds of polystyrene products were marketed in the United States.

These two anecdotes suggest the countless number of accidents that have *not,* at least when they occurred, led to significant discoveries; that is, they have not been serendipitous.

The discovery of low-density, branched polyethylene resulted from a leak in a pressure vessel and the investigation that followed that observation. The discovery of high-density, linear polyethylene resulted from the use of dirty equipment, or at least equipment that had not been scrupulously cleaned since its last use.

In 1953 Dr. Karl Ziegler was the director of the Max Planck Institute for Coal Research in Muelheim, Germany. He had previously followed up and extended the work of Marvel and Friedrich by mixing lithium alkyls and other organometallic reagents (compounds containing a metal

▼ ***Karl Ziegler (with polyethylene bucket) and associates***

atom bonded to an organic group of atoms) with ethylene, trying to polymerize it at lower pressures than those used by the I.C.I. chemists. These experiments produced ethylene polymers that were not large enough to be useful. One day an experiment did not give a polymer at all, but only a dimer of ethylene; that is, only two molecules of ethylene had been coupled together. This result was puzzling enough to make Ziegler and his co-worker, E. Holzkamp, look for the cause. It took several weeks, but they finally found out that the vessel in which this particular reaction had been performed contained a small amount of a nickel compound that had not been completely removed after an earlier experiment.

This discovery sent Ziegler and his staff of chemists on a systematic investigation of the effects of nickel and various other metals and their compounds on the reactions of ethylene. They found that some other metals behaved like nickel and inhibited the polymerization of ethylene, but (and this was the surprising finding) certain other metal chlorides in combination with organo aluminum compounds made very effective catalysts for the production of polyethylene of very high molecular weight, high melting point, and linear molecular geometry—completely unbranched, as depicted in Figure 26.1.

One of the practical dividends from the higher melting point of the new linear polyethylene was that it could be used for tumblers and other kitchenware, which would not melt in the automatic dishwater. About eight billion pounds of high-density polyethylene was produced in the United States in 1987.

Of great importance to the manufacturers was the fact that the polymerization could be carried out at ordinary pressure and low temperature, in contrast to the high pressures and temperatures necessary for the I.C.I. process. The optimum catalyst contained titanium tetrachloride and aluminum triethyl. Because the catalyst was not used up in the polymerization process and could be reused, it did not matter that titanium was expensive. This process became known as the Muelheim atmospheric polyethylene process, and it rapidly received worldwide attention.

Professor Giulio Natta, an Italian chemist who was a consultant for the Montecatini Company of Milan, had immediate access to information about the new polyethylene process by virtue of a licensing agreement between Ziegler's Institute and Montecatini. He applied the Ziegler catalyst to polymerization of propylene and other hyrocarbon relatives of ethylene and showed that it worked for them, too. Thus polypropylene could be produced in high density, high-melting, linear forms that were

▼ *Giulio Natta*

more useful than any previous ones. It has become one of the world's major plastics; nearly seven billion pounds were produced in the United States in 1987. Many molded objects are made of it, including automobile parts, ice chests, and various appliances. Textile fibers for carpets, ropes, and cables are also made of polypropylene. Co-polymers of ethylene and propylene are elastic and can be used in place of rubber in some applications.

Using the new catalysts, for the first time it became possible to make synthetic rubber identical to natural rubber. The new catalyst allowed control of the way in which the monomer units (isoprene) were joined together to give a spiral shape to the polymeric rubber molecule; this is the molecular structure that produces elasticity in rubber. (See the discussion of synthetic rubber in Chapter 11.)

The composition of automobile tires now depends largely on the relative price of natural and synthetic rubber and mixtures are often used. An indication of the impact that Ziegler's discovery had on the polymer industry is given by an estimate of the value of all materials produced by processes based on it as amounting to over a billion dollars per year.

Ziegler and Natta shared the Nobel Prize for Chemistry in 1963. in an international symposium honoring Ziegler shortly after his death in 1973, G. Wilke, his successor at the Max Planck Institute, said: "Karl

Ziegler's guiding principle was that it is not possible to anticipate something which is really new; this can only be discovered by experiment. . . . In addition it was one of Ziegler's principles to keep an eye open for unexpected developments and not to neglect new phenomena as irrelevant to the main project."

▼ POSTSCRIPTS

Poly(vinyl chloride.) Poly(vinyl chloride) (PVC) is a close relative of polyethylene. The monomer vinyl chloride ($CH_2{=}CHCl$) differs from ethylene only in having a chlorine atom in place of one of the hydrogen atoms of ethylene. This polymer is probably the most widely used and industrially important plastic material. Sales in 1987 amounted to over $600 million. It is also the cheapest and probably the most versatile plastic, being used for pipe and pipe fittings (the largest scale use), floppy computer disks, garden hose, building sidings, wire and cable insulation, food packaging, automobile seat covers, shower curtains, and many other household uses. It was among the first plastics to be produced commercially in the early 1930s, and it was probably the first synthetic polymer.

It was first produced entirely by accident. In 1838 a French chemist, Victor Regnault, described the formation of a white powder when sealed glass tubes of liquid vinyl chloride were exposed to sunlight. Much later, in 1872, E. Bauman reported the conversion of vinyl chloride to a white solid mass "unaffected by solvents or acids." The significance of this observation was not recognized, however, until much later, and the commercial development occurred only after practical catalytic methods of polymerizing vinyl chloride were developed.

Poly(ethylene oxide) and Poly(propylene oxide.) In the summer of 1951 George Fowler and Walt Denison, two chemists at Union Carbide Corporation in South Charleston, West Virginia, opened the valve on a tank of ethylene oxide (CH_2CH_2O), expecting the colorless gas to come out. Instead, out came a viscous black liquid. Putting the bad tank aside temporarily, Fowler and Denison sent for another tank and proceeded with their planned experiment.

Fortunately, however, the two chemists were curious enough to investigate the bad tank and its contents. (This is reminiscent of the experience of Roy Plunkett, who discovered Teflon under similar circumstances; see Chapter 27.) Allowing the black liquid to stand in the open, they found it turned into a tough, black solid which, however, was soluble in water. When they filtered the dark solution, it separated into a black powder (on the filter) and a colorless solution. When the clear solution was evaporated, a tough, white solid was left. Analysis of the black solid showed it to contain iron. This gave a clue to the formation of the water-soluble solid. It was a polymer of ethylene oxide and had been formed by the action of an iron oxide catalyst (rust) inside the bad tank. A little experimentation led to procedures for creating polymers of ethylene oxide with molecular weights varying from one hundred thousand to five million. A few years later Charles C. Price would polymerize optically active propylene oxide and produce optically active polymer.

Actually, German chemists had polymerized ethylene oxide 18 years earlier, but no one had noticed the published account. Several properties of the rediscovered poly(ethylene oxide) made it interesting to industrial chemists. One was that even in minute amounts it increased the viscosity of water significantly and made the solutions feel silky. This property was put to use in the thickening of water-based paints and in the formulation of cosmetics, lotions, and liquid dishwashing detergents.

Another use of the water-soluble polymer was in agriculture. Tubular strips could be made and filled with seeds at regular intervals and used in planting. The "seed tape" could be laid in a shallow furrow and covered loosely with soil. When watered by rain or irrigation, the tape would dissolve and the seeds, upon germination, would produce plants at regularly spaced intervals, simplifying cultivation and harvesting.

One of the most fascinating properties of the polymer is that addition of minute amounts to water causes a significant reduction in the friction of water flowing through pipes or hoses. This has been applied in pumping concrete mix through pipes at building sites. It has been used in fire fighting, allowing water even when pumped through smaller hoses to be projected farther with the same power of pumping.

There are several significant "what ifs" in this discovery. For example, what if the polymerization had proceeded further in the bad tank? Nothing would have come out, and the chemists might not have investigated to find out why. Fortuitously, the polymerization had gone just far enough to allow an interesting viscous material to emerge and capture the curiosity and imagination of Fowler and Denison.

27

TEFLON: Out of the Atom Bomb and into the Frying Pan

Teflon is the trade name for polytetrafluoroethylene, the billion-dollar product of the Du Pont company used for various items, from nonstick frypans to space suits to artificial heart valves. Its discovery resulted from an accident observed by Roy J. Plunkett, a young Du Pont chemist who had received his Ph.D. from Ohio State just two years before the fateful day of April 6, 1938. On this day Dr. Plunkett opened a tank of gaseous tetrafluoroethylene in hopes of preparing a nontoxic refrigerant from it. But no gas came out, to the surprise of Plunkett and his assistant, Jack Rebok. Plunkett could not understand this, because the weight of the tank indicated that it should be full of the gaseous fluorocarbon.

Instead of discarding this tank and getting another to continue his refrigerant research, Plunkett decided to satisfy his curiosity about the "empty" tank. Having determined that the valve was not faulty by running a wire through its opening, he sawed the tank open and looked inside. There he found a waxy white powder and, as a chemist, he realized what this must mean.

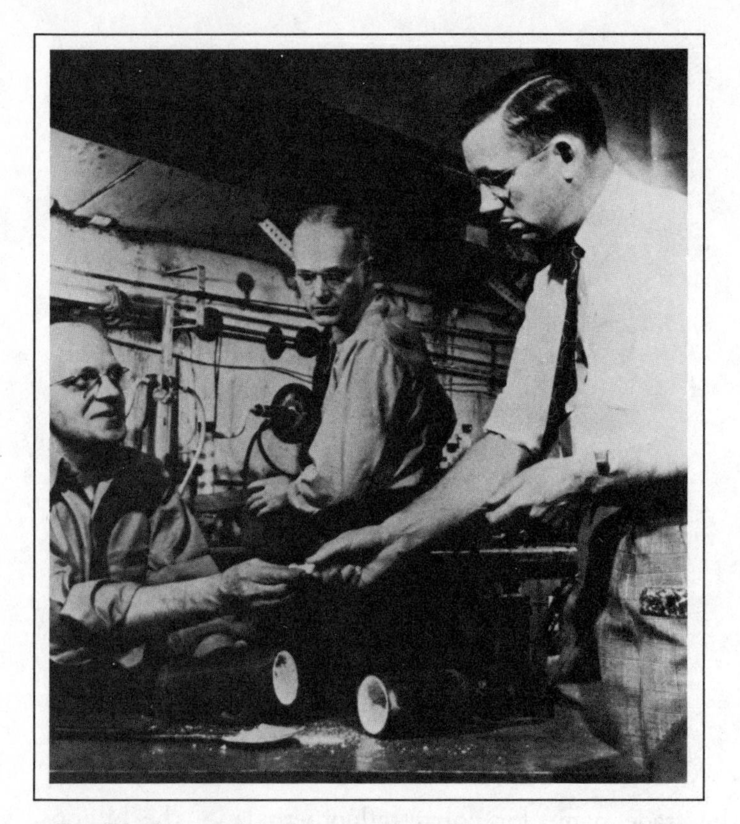

▼ *Roy Plunkett (right), the original "Teflon man," re-enacts the discovery of Teflon in his laboratory*

The molecules of the gaseous tetrafluoroethylene had combined with one another ("polymerized") to such an extent that they now formed a solid material. No one had ever observed polymerization of this particular compound before, but somehow it had occurred inside the mysterious "empty" tank. Encouraged by his chance discovery and by the unusual properties of the polymer, Plunkett and other Du Pont chemists soon found ways to produce "polytetrafluoroethylene" on demand.

The waxy white powder did indeed have remarkable properties: it was more inert than sand—not affected by strong acids, bases, or heat, and no solvent could dissolve it—but, in contrast to sand, it was extremely "slippery." In spite of these interesting and unusual properties, if it had not been for World War II, probably nothing more would have been done with this new polymer for a long time, because it was expensive. In a few months, however, scientists involved in creating the first atomic bomb needed a material for gaskets that would resist the viciously corrosive gas, uranium hexafluoride, one of the materials used to produce the U-235 for the bomb.

General Leslie R. Groves, responsible for the U.S. Army's part in the atomic bomb project, happened to learn from Du Pont acquaintances about their incredibly inert new plastic. (A similar incident of coincidental communication that promoted the wartime development of polyethylene for use as an insulator for radar equipment in England is mentioned in Chapter 26.) Told that it was expensive, General Groves replied that in this project, cost was no object. So the slippery polymer was compounded into gaskets and valves; they were indeed resistant to the corrosive uranium compound. The Du Pont company produced Teflon for this use during the war. The public knew nothing about the new polymer until after the war.

In fact, it wasn't until 1960 that the first Teflon-coated muffin tins and frying pans appeared on the market. Like several new polymer products when they were first introduced to the public, these Teflon products were somewhat disappointing. Although the plastic was ideally suited as a nonstick cooking surface, it was difficult to bond to metal utensils and was, therefore, not resistant to the scouring that housewives were accustomed to give their pots and pans. After various techniques were tried and four generations of Teflon coatings were produced, Du Pont announced in 1986 that their Silver Stone Supra was twice as durable as the third generation Silver Stone. Meanwhile, many other applications had been discovered that made the coating of cooking utensils seem relatively unimportant.

▼ ***Roy Plunkett with cable insulated with Teflon and a muffin tin coated with Teflon***

Roy J. Plunkett was born in New Carlisle, Ohio, in 1910. He received a B.A. from Manchester College in 1932 and, unable to find a job in the midst of the depression, he went on to graduate study at Ohio State University. Both in college and in graduate school, he was a classmate and friend of another famous chemist, Paul Flory. Paul J. Flory was a Nobel laureate in 1974 for his accomplishments in the physical chemistry of polymers. After receiving his Ph.D. from Ohio State in 1936, Roy Plunkett joined the Du Pont company at their Jackson Laboratories, where he was assigned to research on fluorocarbons as refrigerants. During this work the young chemist made his dramatic discovery of Teflon. The development of Teflon was turned over to other divisions in the Du Pont company, which had a major background in polymer products.

Plunkett went on to other work as a chemist and subsequently moved up the managerial ladder in fluorocarbon and tetraethyl lead operations of the Du Pont company. While a manager in Du Pont's Freon Products Division he was instrumental in installing a plant near Corpus Christi, Texas. When he retired from Du Pont in 1975, it was to an island home in Corpus Christi, where he and his wife enjoy fishing and golfing. He has been awarded honorary doctoral degrees from his alma maters, Manchester College and Ohio State, and from Washington College. Further honors include the John Scott medal from the city of Philadelphia, awards from the National Association of Manufacturers, the Society of the Plastics Industry, and the American Institute of Chemists. In 1973 he was inducted into the Plastics Hall of Fame and in 1985 into the National Inventors Hall of Fame.

Even more than these honors, however, he takes pride in the multitude of ways that Teflon has touched the lives of millions of people worldwide. He has said he "can't get over" the letters and calls he still receives from people who are alive today only because a Teflon aorta or pacemaker was placed in their body. Because it is one of the few substances that the body doesn't reject, Teflon can be used for artificial corneas, substitute bones for chin, nose, skull, hip and knee joints, ear parts, artificial tracheas, heart valves, and tendons, sutures, dentures and bile ducts.

Teflon has been used for the outer skin of space suits. It is the insulating material for electrical wires and cables that have resisted the violent radiation of the sun on the moon. The nose cones and other heat shields and the fuel tanks of space vehicles have been fabricated from Teflon.

All of these remarkable and valuable applications have grown out of the serendipitous discovery by Roy Plunkett. An accident, yes, but a *discovery* only because of the curiosity and the intelligence of the man to whom the accident occurred.

▼ POSTSCRIPT

Roy Plunkett was given the assignment of developing a refrigerant from a compound of fluorine because of another serendipitous incident. In 1928 Charles F. "Boss" Kettering, of the General Motors Frigidaire Division, initiated a search for a safe refrigerant—one that would be colorless, odorless, tasteless, nontoxic, and nonflammable, to replace the toxic and noxious substances such as ammonia and sulfur dioxide that were used in refrigerators at the time. After a careful survey of the chemical literature, Thomas Midgely and Albert Henne decided that certain compounds of carbon containing both fluorine and chlorine might be suitable, although compounds of fluorine had sometimes been reported to be toxic.

To test these reports of toxicity, Midgely and Henne needed to prepare samples of a simple chlorofluorocarbon and use them in animal tests. They ordered five one-ounce bottles of antimony trifluoride (the entire supply of this chemical in the USA!) from a chemical supply house. They chose one of the five bottles at random and used it to prepare the chlorofluorocarbon. They placed a guinea pig under a bell jar with the gaseous chlorofluorocarbon compound and found the animal to be totally unaffected by the gas. This confirmed their conviction that organic compounds containing fluorine are not toxic.

As a check on this experiment, the chemists prepared other samples of the chlorofluorocarbon gas using the other bottles of antimony trifluoride, and then they repeated the tests with guinea pigs. In every one of these subsequent tests the guinea pig died! Careful examination showed that all but one of the five bottles of antimony trifluoride contained water. The water in four of the five samples led to the production of the deadly gas phosgene. (The chlorine of the phosgene came from the organic chloride used along with the antimony trifluoride as starting material for the preparation of the chlorofluorocarbon.) It was the phosgene and not the chlorofluorocarbon gas that killed the animals.

If Midgely and Henne had not by chance chosen to use for their first animal test the one bottle of dry antimony trifluoride, they probably would have given up their idea of using chlorofluorocarbons as refrigerants and would have accepted the (incorrect) previous report that these compounds were indeed poisonous. Instead, a joint venture between Frigidaire and Du Pont led to the creation of the Freon Division within Du Pont for research and development of chlorofluorocarbon chemistry, and there Roy Plunkett discovered Teflon.

▼▼▼▼▼▼▼▼▼▼▼▼▼▼▼▼▼▼▼▼▼▼▼▼▼▼▼

28 ▼

Gasoline Technology: Flowery Theories and Gas to Gasoline

▼ Two accidental discoveries in gasoline technology are described in this chapter, one from the early days of the automobile and the other a recent development.

Ethyl Fluid

The search for a gasoline additive to prevent knocking began when Charles F. ("Boss") Kettering became annoyed by the knocking in the 1912–1916 Cadillac engines. The first success resulted from an odd combination of a misconception and an accident. Kettering and Thomas Midgley, a research associate in the Delco company (later absorbed into General Motors), thought that the knock was a delayed explosion due to incomplete combustion of gasoline. They drew an analogy between the early blooming of the trailing arbutus with its rust-colored leaves in the spring, even while snow is still on the ground, and the late-blooming combustion, and hoped that if the gasoline were colored a deep red, it would absorb radiant energy more quickly and vaporize early enough to prevent the knock. So one day in December 1916, Midgley went to the chemical laboratory to obtain some red dye. He found none, but found a bottle of iodine, which was tested because it gave a red color to the gasoline.

Much to the delight of the researchers, the iodine solved the knock problem. But subsequent tests showed that the color of the fuel had nothing to do with the lack of knocking. Nevertheless, the researchers now knew that an additive could somehow prevent knocking. Iodine was too expensive and corrosive to be the answer, so the search went on. After many trials and failures, in 1921 tetraethyl lead was discovered through a conception that was a little more scientific than the trailing arbutus analogy. This conception was based partly on experimental data obtained "Edisonially" (that is, by trial and error, taking every bottle from the shelf and testing its contents), but also partly, more scientifically, on the basis of Mendeleef's theory of the periodic properties of the elements.

The search for a gasoline additive that would prevent knocking began with some pseudoserendipitous discoveries and ended with the planned discovery of tetraethyl lead, which, when dissolved in gasoline, constitutes ethyl fluid. Ethyl fluid was found to be a marvelous antiknock agent, and has been the principal gasoline additive for over 60 years. Now, however, it is a matter of concern because of the danger lead poses in the environment.

From Methane to Gasoline

In 1986 a plant on New Zealand's North Island began producing gasoline from natural gas. With its successful operation, New Zealand has now attained about 50% self-sufficiency in liquid fuels.

The chemistry of the first step of the production is the conversion of methane (CH_4), which is the major component of natural gas, to methanol (CH_3OH, also known as methyl alcohol or wood alcohol). The second step is conversion of methanol to gasoline, which is a mixture of hydrocarbons with molecular sizes in the range C_6 to C_{12} and a volatility in the proper range for automobile engines. This step appears to be more complicated than the first, which is just to add one atom of oxygen to a molecule of methane, but surprisingly, it is not. Indeed, the long-term success of the process might depend on improvements in this first step to reduce its cost, and such improvements may well come from finding an industrial catalyst that can duplicate an enzymatic conversion known to be used by bacteria.

The success of the second step is the direct result of a serendipitous discovery by research chemists at Mobil Oil Company. The June 22, 1987 issue of *Chemical and Engineering News* featured a cover story on the new methane-to-gasoline plant in New Zealand. In a letter to the same journal in the September 2, 1987 issue, William H. Lang, a retired research chemist, recalled that discovery. He wrote that on March 10, 1972 he and Clarence Chang were trying to make neopentane from isobutane and methanol using a crystalline silica-alumina catalyst labeled ZSM-5 [isobutane has the formula C_4H_{10}].

> All of the methanol and part of the isobutane were converted to liquid hydrocarbons but no detectable neopentane. Having previously conducted research in the reforming of naphtha (a gasoline-size hydrocarbon mixture), I immediately recognized the composition of the hydrocarbon products as high-octane gasoline. Our next experiment with methanol and the same catalyst confirmed the conversion of methane to gasoline-range hydrocarbons. . . . Many years of research and development have brought this original discovery into a successful process in New Zealand.

Dr. Lang's mind was prepared through previous research in gasoline chemistry and so he was able to make the most of the accidental observation.

29

DRUGS Accidentally Found Good for Something Else

Penicillin, sulfa drugs, cephalosporins, and cyclosporine were discovered by accident. *Most* drugs have been discovered through serendipity or at least through pseudoserendipity. A drug being used for one purpose has often been found effective for an entirely different and sometimes more significant purpose. This chapter gives a sampling of some of the discoveries of this type.

Aspirin

Aspirin has been one of our most widely used drugs for many years; it has recently acquired even more respect. It was first prepared for use as an internal antiseptic, but it was found ineffective. However, it was found to be a valuable analgesic and antifebrile (fever-relieving) drug, and now it is being recommended for preventing heart attacks.

Soon after Joseph Lister pioneered the use of phenol as an antiseptic in surgical operations, scientists thought of finding a drug that could be administered internally to patients suffering from bacterial diseases. In the 1870s salicylic acid was synthesized and used because it was known to yield phenol in the body; however, although it reduced fever, it did not affect the infection causing it. It also produced nausea. Felix Hoffmann, a chemist at the Bayer company, later prepared a modified form of salicylic acid, the acetyl derivative, which was found to be effective against fever and the pain of arthritis, and had fewer undesirable side effects. The name *aspirin* came from the fact that salicylic acid had originally been obtained from *Spiraea* plants; the prefixed *a* was added to signify acetyl.

Since its entrance into the pharmaceutical market in the 1890s, more people have used aspirin than any other drug. Its potential value in preventing heart attacks is a development that is not yet fully tested, but already more than forty million pounds are being produced in the United States annually, amounting to approximately 300 tablets per person each year!

Psychoactive Drugs

Before the 1950s, psychiatric illness was untreatable by drugs. Schizophrenic persons and persons who suffered from severe manic depressive illness and neuroses had to be confined to mental hospitals. Then, within about 10 years, treatment with psychotherapeutic drugs provided hope for a more normal life, and the mental hospitals were virtually emptied. There was serendipity in the discovery of nearly every psychoactive drug.

Chlorpromazine. In the latter part of the 1940s, the French neurosurgeon Henri Laborit wanted a drug to calm his patients before an operation—prior even to anesthesia. He thought that an antihistamine might serve this purpose, because histamine was known to be released in the patient's body by anxiety preceding the operation. A pharmaceutical company supplied Dr. Laborit with a sedating antihistamine called promethazine. It seemed to help some, and so the surgeon asked for a stronger antihistamine. The second drug given was chlorpromazine. Laborit was so impressed with the "euphoric quietude" produced in his patients before their operations that he began recommending the use of this drug to his colleagues.

Two other French doctors, psychiatrists Jean Delay and Pierre Deniker, found not only that the drug was useful just before operations on normal surgery patients, but that it was effective in calming their patients in the manic stage of manic-depressive illness. They, as well as other psychiatrists, tried it on other mentally ill patients between 1952 and 1955, and they found that it was especially effective in the treatment of schizophrenia.

By the late 1950s, chlorpromazine began to be used widely in Europe and the United States. Within a decade severely ill schizophrenics could be discharged from asylums, where they had been kept in padded cells and straitjackets, to return to productive jobs and almost normal lives in the community. In the United States the population of state mental hospitals declined by hundreds of thousands; community health movements, aimed at maintaining mental patients at home with family and friends, became possible only because of the ability of chlorpromazine to ease the psychotic symptoms of even the most severely disturbed patients.

The use of chlorpromazine initiated a second revolution in medical treatment of the mentally ill. Clinical tests found that excessive dosage of the drug caused symptoms similar to those of Parkinson's disease. This finding led to studies of the effects of drugs on different areas of the brain and of molecular abnormalities of the brain associated with mental illness.

Dopamine, a neurotransmitter that occurs in high concentrations in the parts of the brain that regulate motor activity, was found to be almost totally absent in the brains of patients who had died with Parkinson's disease. Injections of L-dopa, a chemical that is transformed by natural processes in the brain into dopamine, gave dramatic improvement to patients with Parkinson's disease.

Arvid Carlsson, a Swedish pharmacologist, speculated that drugs such as chlorpromazine, while controlling schizophrenia, might actually create the effects of a deficiency of dopamine in the brain by blocking dopamine receptors. In 1975 when it became possible to measure receptors for dopamine in the brain, in the Johns Hopkins laboratories of Dr. Solomon H. Snyder, the therapeutic effects of antischizophrenic drugs such as chlorpromazine were proven effective because of their blockage of dopamine receptors. These findings suggested that what is fundamentally wrong in the brains of schizophrenics is either an excess of dopamine formation or a supersensitivity of dopamine receptors.

Imipramine. The serendipitous success of chlorpromazine against schizophrenia stimulated the synthesis and testing of a host of chemically similar drugs. One of these was imipramine, synthesized by the Swiss pharmaceutical firm Geigy. When Dr. Roland Kuhn found it to be surprisingly inactive against schizophrenia, he proceeded to evaluate it in other types of mental illness. In 1957 he reported its impressive activity as an antidepressant, an almost opposite effect from that of chlorpromazine!

Lithium. The discovery of lithium as a psychoactive drug was the most improbable of all. Late in the 1940s, John Cade, a young Australian psychiatrist, speculated that the mania associated with manic-depressive illness might be caused by the abnormal metabolism of uric acid. To test this theory, he injected uric acid—in the form of a lithium salt, and along with it lithium carbonate—into test animals and observed dramatic therapeutic responses. Although he published his findings in an Australian journal, few psychiatrists took note of his observations until the mid-1950s.

Then a Danish doctor, Mogens Schou, happened to read Cade's paper. He tested Cade's compounds for the treatment of mania and found them effective. However, it soon became apparent that the uric acid part of the drug had nothing to do with its effectiveness. It was only the fact that a *lithium salt* of the acid had been used that was responsible for the therapeutic effect; other lithium salts were found to be equally good.

Because lithium salts are common and could not be patented as a drug, pharmaceutical companies were reluctant to commit to large-scale production for clinical use. Another factor that delayed the clinical use of lithium was the anticipated danger that the lithium ions, in large doses of the drug, might compete with the chemically related sodium ions in the body and have toxic effects.

Thus it was not until 1970, more than 20 years after the discovery of the value of a lithium salt in controlling manic-depressive illness, that lithium was introduced into American psychiatric practice. The simple lithium ion is the most effective agent ever identified for the treatment of mania, but the mode of its action is still a mystery.

Librium and Valium. The year 1960 was an eventful one in the history of chemotherapy. Two drugs launched that year were to have a momentous impact on society in different ways: Librium, the first of a new series of tranquilizers, and an estrogen oral contraceptive. (The discovery of the birth control pill is described in Chapter 20.) Librium, the most prescribed drug in the 1960s, was replaced in the top position in the 1970s by its chemically related stablemate Valium. The story of the development of these two tranquilizers is one of pseudoserendipity. Leo

Sternbach, of the Hoffmann–La Roche company, set out to find an antianxiety drug.

Leo Sternbach was born in 1908 in a town in the Pula peninsula, at that time a part of Austria–Hungary, but now Yugoslavia. He took master's and doctor's degrees in pharmacy at the University of Krakow, which is now in Poland. In 1940 he was working for Hoffmann–La Roche in Basel, Switzerland; the next year he transferred to their laboratory in Nutley, New Jersey. In 1946 he became an American citizen. He retired as director of medicinal chemistry for Hoffmann–La Roche in 1973.

In 1953 Dr. Sternbach was assigned to find a new class of anxiolytic (antianxiety, or tranquilizer) drugs. The desired compounds had to offer some synthetic challenges and be relatively unexplored, to reduce the risk that competitors would come up with the same drugs. However, economic considerations required that the compounds be readily prepared from accessible starting materials. In addition, it would be desirable that the basic molecular unit of the compounds allow conversion into a large number of derivatives to increase the chance of finding a potent drug. Sternbach thought back to his investigations in Krakow of compounds that were possible starting materials for the synthesis of dyes.

They fulfilled most of the criteria of desirable anxiolytic drugs. They were obscure (no one had done anything with them in the 18 years since Sternbach had prepared them), they could be made rather easily, and they had a nitrogen atom attached to an aromatic ring like several compounds of known biological activity. There turned out to be one major problem, however: they just did not have the desired tranquilizing properties.

In April 1957 Sternbach told his research group to terminate the work with these compounds and go on to other types. During the cleanup, one of his chemists showed him a compound that they had prepared two years before, but had not tested pharmacologically. They tested it thinking that a negative result from this compound would confirm that this type of chemical was hopeless and they could turn their attention to other types.

Much to Sternbach's surprise, this overlooked compound showed powerful tranquilizing effects. When his chemists sought an explanation for the anomalous properties of this compound (which they had thought was chemically similar to the compounds that had no tranquilizing effects), they discovered that this compound was not what they thought it was. During its synthesis, it had undergone an unexpected intramolecular rearrangement, to produce a quite different structure. They then synthesized other compounds analogous to the true structure of the forgotten compound, but none was as good as the original.

After further careful clinical tests, the "forgotten compound" was put on the market in 1960 as Librium, and it was an immediate success. Even before Librium was marketed, however, the Hoffman–La Roche chemists had found that a hydrolysis product of Librium, which they thought was produced in the body from the drug, was equally potent. They synthesized various derivatives of the hydrolysis product and found one of them was superior in several ways to Librium. The new drug was named Valium. Put on the market in 1963, it soon replaced Librium in usefulness as a tranquilizer.

Surprisingly, chlorpromazine, a powerful tranquilizing antischizophrenic drug, and imipramine, an antidepressant drug—that is, drugs that have almost opposite pharmacological effects—have similar molecular formulas. The formulas of Librium and Valium are similar, but not identical. The accidental discovery of LSD, a highly psychoactive substance, although not useful as a drug, is described in Chapter 20.

Since the introduction of Valium, 12 other drugs with similar chemical structures have been developed and sold in the United States, and a further 21 in other countries around the world, often as a result of less stringent regulations, but no fundamental pharmacological advances in tranquilizers have occurred since the accidental discoveries of Librium and Valium.

Antiarrhythmic Drugs: So Things Won't Go Bump in the Night

Novocaine and Xylocaine (generic names, procaine and lidocaine) have been widely used as local anesthetics (for example, in dentistry). In the 1940s it was discovered accidentally that injection of novocaine into dogs that had developed life-threatening cardiac arrhythmias (irregular heartbeats) when placed under general anesthesia restored their hearts to normal beating. Soon after this became known to anesthesiologists, opportunity was afforded to evaluate this effect in human beings in one of the U.S. Army Chest Centers, on patients who had acute cardiovascular dysfunction. These pioneering tests were performed by Major Charles Burstein and reported in 1946.

One must assume that these original tests on humans were done in cases where the "dysfunction" was extremely dangerous. Injection of novocaine into a conscious patient had always been scrupulously avoided, since it was known that convulsions could thus be produced. In cases where the local anesthetic was injected into an *anesthetized* patient undergoing severe arrhythmias, however, no adverse effects were seen and, in fact, cardiocirculatory improvement was observed. In 1950 J. L. Southworth and his collaborators presented clinical evidence that Xylocaine had a good antiarrhythmic effect. Since then the use of intravenous injection of various local anesthetics in connection with cardiac surgery has become common practice.

This represents another clear example of a drug developed for one purpose being accidentally found to be useful for a quite different one.

Minoxidil: A Hair-Raising Experience

In 1980 Dr. Anthony Zappacosta of Bryn Mawr, Pennsylvania, wrote a letter to the *New England Journal of Medicine* describing the new growth of hair on the scalp of a 38-year-old male who was being treated for high blood pressure with a drug called minoxidil. The patient had been nearly bald since the age of 20. Zappacosta's letter and other reports that began to appear aroused a great deal of interest among dermatologists—and among the general male public, of course, as soon as the news reached them. After all, hair loss has been a cause of concern and a cure has been sought since the time of ancient Egypt.

One of the dermatologists was Virginia Fiedler-Weiss of the University of Illinois Medical Center in Chicago. She wondered whether the drug would promote hair growth without the powerful effect on blood pressure and without undesirable side effects if it were applied topically. She prepared a lotion from crushed minoxidil tablets and had three patients apply it to their scalps twice a day. Within weeks she found two of the patients growing hair! When Dr. Fiedler-Weiss called Upjohn, the maker of minoxidil, she learned that company had already started preliminary studies using a solution of the drug. Fiedler-Weiss later collaborated with Upjohn on a larger study. A paper published in 1984 reported that 25 of 48 patients grew new hair, but only 11 patients had cosmetically acceptable results.

By 1987 many other studies had been made. The conclusions reached may be stated as follows: Minoxidil solution is unlikely to help most people with severe hair loss. The drug's best use will be to help patients hold on to the hair they already have and to help reverse early baldness. It seems to work only if a person uses it indefinitely, and the safety of long-term use on the skin is uncertain. In late 1987 it had not been approved for general use by the FDA, and even if it were, it would require a prescription.

Nevertheless, minoxidil, a drug used to combat high blood pressure, has been proven by clinical trials to produce new growth of hair. A paper in *Clinical Pharmacy* (May 1987) concluded that "prognostic indicators as to which patients will respond to therapy are not yet adequately defined. Perhaps formulation of an optimal vehicle would enhance efficacy, but the potential for increased adverse reactions must be considered. More information is needed before the place of minoxidil in hair loss treatment can be established."

Interferon: Cancer and Arthritis

In 1955 Isaacs and Lindenman described interferons as a group of proteins and glycoproteins, produced by cells in response to virus infection, which could inhibit the growth of a wide range of viruses. The possibility of a new type of cancer drug caught the imagination of the press, the public, and the medical world.

However, progress in applying these agents in the clinic was very slow because it was difficult to obtain adequate amounts of pure material. The situation changed dramatically a few years ago when the first human interferon genes were cloned.

A chance observation during a clinical cancer experiment in 1984 led to the surprising discovery that injections of interferon seemed to relieve the pain and swelling of rheumatoid arthritis. Preliminary studies showed that the hormone worked in about two-thirds of people who are not helped by conventional treatment. If these results prove to be general, it could provide a large, unexpected new market for genetic engineering's mass-produced human hormones.

The serendipitous discovery of the effect of the cancer drug against arthritis was made during a study at Bioferon, a German subsidiary of the biotechnology firm Biogen. Dr. Seth Rudnick said that a few patients who had both cancer and rheumatoid arthritis showed some improvement in the pain from arthritis. Company officials were skeptical at first, but they tested 38 patients in Germany and, within a week or two of the injections (given five times a week), 28 found easing of the pain of arthritis and the pain even disappeared for some. Further testing must be done to compare interferon with dummy placebos, and attempts must be made to learn whether the hormone affects the underlying disease of arthritis as well as relieving its symptoms.

▼▼▼▼▼▼▼▼▼▼▼▼▼▼▼▼▼▼▼▼▼▼▼▼

30 ▼

Drugs from Sewage and Dirt

▼ Two of our most important modern drugs were discovered in sewage and dirt! The antibiotics known as cephalosporins, representing a new type of penicillin, were found among the sewage from Cagliari, Sardinia. Cyclosporine, the antibiotic that facilitates human organ implants, was found in soil from Wisconsin and Norway that employees of a pharmaceutical company carried back to Switzerland. In both cases fungi produced the antibiotics.

To most people the word *fungus* brings to mind unpleasant things like athlete's foot or the mold on bread and shower curtains. To the farmer they are more than unpleasant, because they can be devastating to crops. But fungi have good aspects: the production of beer and wine, the raising of bread, and all the industrial operations involving fermentation depend on yeasts, which are fungi.

One of the most important functions of fungi is the production of antibiotic compounds, poisonous to competing organisms, especially bacteria.

Cephalosporins

In the early 1950s doctors using penicillin encountered a serious problem: a number of bacteria, especially staphylococci, were found to be unaffected by penicillin. A key to the solution of this problem had been discovered several years earlier, but it was not known for some time.

In 1948 Giuseppe Brotzu, a professor of bacteriology in Sardinia, the large Italian island in the Mediterranean Sea, isolated an antibiotic substance produced by a fungus, a strain of *Cephalosporium acremonium*, in the sea near a sewage outfall from the city of Cagliari. Professor Brotzu wanted to study the possible function of the antibiotic in the self-purification of sewage. Finding the concentrated culture fluid from the fungus an effective antibiotic, he by-passed animal tests and administered the material to human patients with infections, both by local application and by injection. He concluded that his patients, especially those with typhoid fever, showed some improvement.

In 1948, after failing to interest any Italian pharmaceutical company in his findings, he published them in a journal titled (in Italian) *Works of the Institute of Hygiene of Cagliari,* stating that he hoped others would take up this work, because he had only limited facilities and equipment in Sardinia. He also reported his findings to a former British public health officer in Sardinia, who brought them to the attention of Sir Edward Abraham at the School of Pathology at the University of Oxford, in England. Years later, Abraham wrote: "Had it not been for these events it is unlikely that we should have heard of Brotzu's work. We thought at first that his publication comprised an issue of a local journal. However, when I asked him later how often the journal was published he replied with a smile that it had never been published before and had never appeared since, but that there would be a further number if he again found anything of comparable interest."

Abraham initiated a concentrated study at Oxford of the antibiotic substances produced by the *Cephalosporium.* The Oxford scientists found that the fungus produced several different antibiotics. The first one isolated was active mainly against gram-positive bacteria, so it was named cephalosporin P. Brotzu, however, had noted that his preparations from the fungus were active against both gram-positive and gram-negative bacteria, so the Oxford workers tried to isolate other antibiotics from the culture medium. By 1954 Abraham and his co-workers had obtained an almost pure form of another antibiotic, which they named cephalosporin N because it killed gram-negative bacteria. Further chemical studies showed this substance to be so closely related to penicillin that it was renamed penicillin N.

By 1955 a pure crystalline substance named cephalosporin C was obtained. It had an extremely low toxicity and was effective against some of the virulent bacteria that were resistant to the penicillins in general use in the 1950s. In 1961 Abraham and Guy Newton announced the molecular structure of cephalosporin C, based on chemical evidence. In the same journal issue, Dorothy Crowfoot Hodgkin and E. N. Maslen confirmed the structure on the basis of their X-ray diffraction analysis. Dorothy Hodgkin had earlier provided the compelling X-ray evidence for the structure of penicillin in the 1940s; she was awarded the Nobel Prize in Chemistry in 1965. Although workers at Oxford did most of the isolation and structure elucidation of the cephalosporins, Robert B. Woodward and his co-workers at Harvard University synthesized cephalosporin C, and Woodward announced the synthesis in his Nobel Prize address in 1966.

The yield of cephalosporin C from the original fungus was too low to be practical, but two developments led to the use of cephalosporins in medicine. The first was the finding of a mutant strain of the fungus that produced substantial amounts of cephalosporin C; the second was the discovery of ways to modify the cephalosporin C molecule chemically to produce a wide spectrum of medically valuable β-lactam antibiotics. A β-lactam is a molecule containing a four-membered ring that includes a nitrogen atom. The penicillins are also β-lactams, but have a second *five*-membered ring whereas the cephalosporins have a second *six*-membered ring, as shown in Figure 30.1. Nearly 20 cephalosporins are available today in the United States and many more are on the synthetic horizon. The six newest third-generation cephalosporins all have the same basic β-lactam nucleus and differ in the substituents at the positions indicated

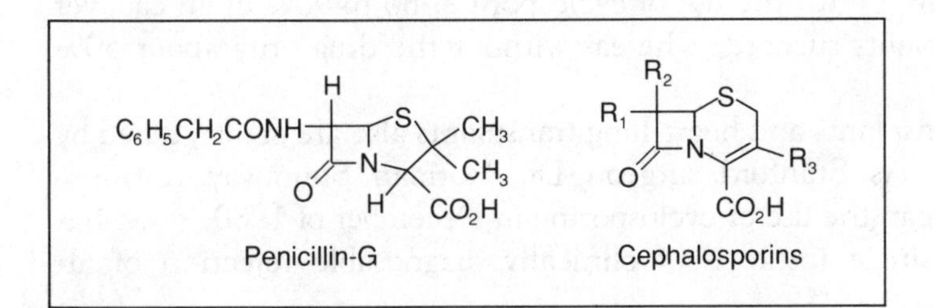

FIGURE 30.1. A specific molecular formula for a penicillin and a general formula for cephalosporins

by R_1, R_2, and R_3 in Figure 30.1. They all have an extended spectrum of gram-negative activity, but with important differences among specific agents.

A strange chain of events followed the initial inquiry into how sewage that had been dumped into the sea from a Mediterranean coastal city became purified in seawater, and led to the discovery of a completely new arsenal of antibiotic drugs!

Cyclosporine

Since 1981 cyclosporine has revolutionized the transplantation of human organs. This drug prevents rejection—the action of the body's immune system when it recognizes foreign tissue and makes transplanted organs inoperable, sometimes almost immediately after transplantation. This drug also has surprisingly few side effects. Dr. Thomas Starzl, a surgeon at the University of Pittsburgh School of Medicine, reported in 1983 during congressional hearings called by Albert Gore, Chairman of the House Committee on Science and Technology, that before this drug became available in 1979 liver transplants were risky at best. "For almost 20 years there seemed no way out of the dilemma. The drugs being used were, on the one hand, unreliable in preventing rejection and, on the other, extremely dangerous." With cyclosporine, however, the percentage of patients whose liver grafts survive for the crucial first year has increased from about 35% to 65 or 70%.

Dr. G. Melville Williams of Johns Hopkins University testified about kidney transplants: "Four centers in this country have had experience with the new immunosuppressive drug cyclosporine and all of these investigators agree that many of the practices we have applied [to increase graft survival] may prove to be obsolete when this drug becomes available for general use." With the use of cyclosporine, 80 to 90% of all cadaver kidney transplants succeed, whereas without the drug only about 50% do.

Heart transplants and heart-lung transplants also are greatly aided by cyclosporine. As Stanford surgeon Dr. Norman Shumway testified, "Since we began the use of cyclosporine in December of 1980, there has not been a single instance of clinically diagnosable rejection of allografted heart."

To understand the discovery of cyclosporine, you must know that some pharmaceutical companies suggest to their employees who travel around the world that they bring back samples of soil to be tested for microorganisms that may be antibiotic. The Sandoz Corporation in Basel, Switzerland, is one such company. In 1970 microbiologist Jean Borel investigated some soil brought from Wisconsin in the United States and from Norway by their travelers. His investigation revealed that the soil from both areas contained two new strains of fungi that produced a water-insoluble substance. This substance, which was named cyclosporin A, did not have strong antibiotic properties, but its unusually low toxicity led to further tests, and Borel made the surprising discovery in January 1972 that it had a marked immunosuppressive effect. Soon after this, however, the management of Sandoz decided to de-emphasize their efforts in immunology, and asked Borel to discontinue the work on cyclosporin A.

Borel protested this decision vigorously and, fortunately, was allowed to continue his research on cyclosporin A. He found that the immunosuppressive effect could be demonstrated in all of the animal species tested. When he turned to tests on humans, he ran into a problem of absorption because of the lack of water solubility of the agent. Human volunteers ingested it in gelatin capsules, but little or no drug could be found in their bloodstreams. At this stage of the study, Borel was convinced that the problem was the method of delivery and that it could be solved; he volunteered to drink a cocktail consisting of the drug dissolved in almost pure alcohol containing a little water and an emulsifying agent. He reported that he "got tipsy," but two hours later a pharmacologically active concentration of the drug could be detected in his blood! A better vehicle of oral delivery was later found to be olive oil.

In June 1978 surgeons in England investigated the first human patients, in mismatched cadaver kidney transplantation and in bone marrow transplantation. Now many years of effort by dozens of scientists have succeeded beyond expectations. The FDA has approved the drug, and it is in general usage in centers where organ transplants can be done. However, an organ must be transferred from a donor body into an acceptor body within 24 hours. This limitation often requires great haste in removing the organ from a body at one site, transferring it (usually by air) to another site, and inserting it into a patient's body after the patient has been prepared for the transplant by medication with cyclosporin A. The approved name for cyclosporin A has now been changed to simply cyclosporine.

Another surprising potential use for cyclosporine is in treating parasitic diseases. The drug kills schistosomes, the worms that cause the tropical disease schistosomiasis. Dr. Ernest Bueding of Johns Hopkins University School of Medicine originally thought that perhaps the drug might ameliorate to some degree the symptoms of the disease. To his amazement, "purely by accident we found [a direct] effect on the worms." Cyclosporine also inhibits the malaria parasite. This also was discovered by accident in experiments on mice, one of the few animals that is susceptible to malaria. It even proved effective against chloroquine-resistant malaria strains. (Chloroquine was effective against malaria in the 1940s, but resistant strains were encountered during the Vietnam war.)

No one yet knows how cyclosporine works, either to prevent rejection or to kill parasites. Novel in chemical structure, it is a cyclic molecule made up of 11 amino acid components. One of these is a previously known compound but in the unusual D-form; most of the natural amino acids are in the L-form ("left-handed form"). (See the story of Pasteur and right- and left-handed molecules in Chapter 12). Another one is completely new in nature. The molecule has been synthesized, and chemists are busy preparing derivatives and analogs to see what portions of its structure are necessary for its biological effects. As yet, none has been found that is superior to cyclosporine.

This discovery is another example of serendipity. Although Sandoz was looking for interesting antibiotics in the soil from different areas of the world, they found a drug that revolutionized the surgical transplantation of vital organs such as hearts, lungs, liver, and kidneys—if not more important than finding another antibiotic, at least something very different from what they expected.

31▼

Brown and Wittig: Boron and Phosphorus in Organic Synthesis

▼ The inscription for the 1979 Nobel Prize in Chemistry reads: "half each to Professor Herbert C. Brown, West Lafayette, Ind., and Professor Georg Wittig, Heidelberg, for their development of boron and phosphorus compounds, respectively, into important reagents in organic synthesis." These two men, from different continents and different backgrounds, were brought together to share the Nobel Prize because of similar contributions to the scientific community. They also shared the blessing of serendipity in their pathways to this recognition.

Hydroboration

Born in London, England, Brown was brought to America in 1914, when only two years old. He spent his early life in Chicago, Illinois, where he received his education, graduating with the first class of Wright Junior College in 1935 and obtaining his B.S. from the University of Chicago in 1936, where he completed two years of study in one year. One of Brown's first encounters with serendipity was unrecognized at that time: for a graduation present his girlfriend (and future wife) gave him a book titled *The Hydrides of Boron and Silicon.* She chose this gift, which was to have a profound influence on the direction of the professional career of Herbert Brown, because it was among the least expensive of those on chemistry in the bookstore, and this was during the great depression. The future Mrs. Brown is said to have inscribed her gift "To my future Nobel laureate," and Brown credited his parents with remarkable foresight in giving him the initials H. C. B. (the chemical symbols for hydrogen, carbon, and boron).

Be that as it may, Brown developed a strong interest in the chemistry of boron and decided to undertake graduate study with Professor H. I. Schlesinger in this area at the University of Chicago, where he was granted a Ph.D. in 1938. As a postdoctoral research fellow he continued to work with Schlesinger on compounds of boron and uranium under the auspices of the Defense Department during World War II. In the course of this work they developed new and practical methods for preparing lithium and sodium borohydrides.

After four years on the faculty at Wayne State University in Detroit, Brown went to Purdue University as professor of inorganic chemistry, where he continued his interest in compounds of boron. The research on sodium borohydride had been done under government contracts during

▼ ***Herbert C. Brown***

the war and the results could not be published for several years after the war because they were classified. The research on lithium aluminum hydride, on the other hand, was done after the war and did not suffer from this difficulty. It was published first and attracted a great deal of attention from the chemical community.

In the 1950s at Purdue, Brown set out upon a comprehensive program of research on the effects of solvents, substituents, and metal ions on the reducing properties of sodium borohydride, the compound he felt had been neglected because of the earlier publicity on lithium aluminum hydride. Sodium borohydride was known to reduce (add hydrogen to) the carbon-oxygen double bonds of compounds known as esters, such as ethyl benzoate and ethyl stearate, when the reaction was catalyzed (speeded up) by the presence of aluminum chloride along with the sodium borohydride. Each of these two esters took up exactly two hydrogen atoms from an equivalent amount of sodium borohydride, but when an analogous experiment was done with ethyl oleate, the amount of sodium borohydride used up by this ester corresponded to significantly more than two atoms of hydrogen.

The chemist working with Professor Brown on this experiment was Dr. B. C. Subba Rao, who had received his Ph.D. from Purdue in 1955 for research carried out under the direction of Professor Brown, and who was now continuing to collaborate with him. When the anomalous behavior of ethyl oleate was brought to the attention of Brown, it was at first credited to impurities in the sample of ethyl oleate. Brown suggested that this explanation should be tested by careful purification of the sample of ethyl oleate and repetition of the experiment with purified material. In *Boranes in Organic Chemistry* (1972), Brown points out that "the research director is in an enviable position to insist on high standards—he does not have to do the actual experimental work! I persuaded Dr. Subba Rao to return to the bench and repeat the experiment with purified ethyl oleate."

However, the repeated experiment gave the same result, showing that it was not an impurity in the sample of ethyl oleate that was responsible for the additional reaction with sodium borohydride and, in fact, the careful repetition of the experiment showed that ethyl oleate was reacting with *three* equivalents of sodium borohydride whereas ethyl stearate required only two. Brown then realized this was evidence that a part of the ethyl oleate molecule other than the carbon-oxygen part it shared with ethyl stearate was reacting with (using up) the sodium borohydride. The difference in ethyl oleate and ethyl stearate is the existence of a *carbon-carbon* double bond in the ethyl oleate molecule that is not present in ethyl stearate; Brown recognized that this bond was responsible for the additional reaction with sodium borohydride. This was entirely unexpected.

The next step was to test a molecule that had a carbon-carbon double bond but no carbon-oxygen double bond (a simple alkene), to see whether it would react with sodium borohydride and aluminum chloride. When it did, the researchers had to determine exactly what was happening, because no such reaction had been observed before.

Brown decided that the reaction must involve the intermediate formation of diborane, produced from sodium borohydride and aluminum chloride, and it was the diborane that was adding to the double bond, giving a product called an organoborane. The addition of diborane to an alkene had been observed before. The reaction was called hydroboration because it involved the addition of hydrogen and boron to another molecule. In previous work, hydroboration occurred only under extreme experimental conditions and to an impractical extent. Brown reasoned that their success in producing hydroboration in good yield under mild conditions must be due to a catalytic effect of the aluminum chloride. He and Subba Rao demonstrated that they could indeed produce diborane from sodium borohydride and aluminum chloride in an ether solution (the ether they had been using was a nonvolatile commercially available solvent, not the ordinary diethyl ether that is an anesthetic) under mild conditions, and that the diborane would add to simple alkenes.

Before publishing a report of this new, practical procedure for hydroboration, the success of which they attributed to the catalytic effect of aluminum chloride, they decided to do one more experiment to confirm their theory: an experiment using diborane with the alkene in the ether solution, but omitting the aluminum chloride. To their surprise, the hydroboration occurred just as well without the aluminum chloride! Further thought and experimentation led to the conclusion that the inadvertent use of an *ether solvent* was the secret to the successful hydroboration.

Professor Brown and his collaborators went on to develop the hydroboration procedure into a rapid, convenient, and general method of preparing organoboron compounds. Once these compounds became readily available, they were enormously useful as intermediates in the synthesis of many valuable organic compounds, most of which do not contain boron. The extent of the usefulness of hydroboration is attested to not only by the Nobel Prize shared by Brown, but also by books, reviews, and research publications by chemists all over the world who have profited from the serendipitous discovery of the "vast unexplored continent" (as Brown called it) of hydroboration.

Brown's discovery provides the ability to synthesize not only complicated organic molecules but also simple ones that cannot be made conveniently any other way. For example, isopropyl alcohol, which can be found at any drugstore as the main component of rubbing alcohol, is prepared simply and cheaply from propylene, a petroleum-based starting material. Its isomer, normal propyl alcohol (*n*-propyl alcohol), however, cannot be made conveniently from propylene because of the principle described by Markovnikov's rule. By using Brown's hydroboration technique, *n*-propyl alcohol can be made from propylene by a procedure that involves "anti-Markovnikov addition" as one step, a technique that is described in every elementary organic textbook now.

▼ POSTSCRIPT

Brown has described another serendipitous discovery about the properties of sodium borohydride. During World War II the U.S. Signal Corps wanted a practical source of field preparation of hydrogen gas. They heard that Schlesinger and Brown had a way to produce the gas from sodium borohydride, and they asked for a good method of preparing pure sodium borohydride. In experiments aimed at developing such a method, Brown tested a purification procedure in which the sodium borohydride would be extracted with acetone, a common solvent. The borohydride was found to dissolve in acetone with the production of heat, and when the solution was examined, it was found that the acetone had been converted into isopropyl alcohol. In this way, the very useful and general procedure of converting aldehydes and ketones (acetone is a ketone) into alcohols by reaction with sodium borohydride was discovered.

Although the approaching end of the war precluded development of field use of sodium borohydride for generation of hydrogen, the discovery of the reduction of the functional group of aldehydes and ketones by sodium borohydride revolutionized the methods used by organic chemists for such reactions, and this compound, "a product developed under the exigencies of war research, later found its main application in the pharmaceutical industry."

Synthesis of Alkenes

Georg Wittig, who shared the Nobel Prize with Brown in 1979, was born in Berlin in 1897. As a young man he displayed talents and interest in both science and music. Another famous person who was similarly torn between science and music was the Russian chemist and composer Alexander Borodin. Many do not realize that Borodin was a teacher of chemistry and medicine, because he is known to most as the composer of the first truly Russian music. Although we cannot know what music the world may have missed because of Wittig's choice of a profession, a host of chemists who have profited from his valuable contributions in both organic and inorganic chemistry can be grateful that music did not win the prime affection of this man. A third love of Wittig as a young man and throughout his life was mountain climbing. Fortunately he was able to follow this calling as a hobby, and some have seen a connection between his mountaineering and his approach to chemical research, in that he explored rare and high areas in both activities.

Wittig's professional career started at the University of Tübingen in 1916. It was interrupted by military service during World War I, but he finished his undergraduate study at the University of Marburg and continued there to receive a Ph.D. in 1926. He taught at Marburg until 1932, then went to the Technical College of Braunschweig, then the University of Freiburg (Germany), and in 1944 he became a full professor and director of the Institute of Chemistry at the University of Tübingen. In 1956 he moved to the University of Heidelberg as head of the department. He became professor emeritus there in 1967 before he shared the Nobel Prize with H. C. Brown in 1979. He died in August 1987 at the age of 90.

During his tenure at Tübingen in 1953, he encountered the serendipity that led to his Nobel Prize. Wittig and his student Georg Geissler were studying compounds in which a phosphorus atom was combined with five other atoms, that is, pentavalent phosphorus compounds. In one experiment they found that the organophosphorus compound reacted with a compound containing a carbon-oxygen double bond (which is a characteristic structural unit in the molecules of the important classes of compounds called aldehydes and ketones) in such a way as to replace the carbon-oxygen double bond with a carbon-carbon double bond. In the original research paper with Geissel, Wittig wrote: "The behavior of V [the organophosphorus compound] with benzophenone [the ketone containing the carbon-oxygen double bond] was astonishing."

Although this was a small part of this research paper and was obviously unexpected, the fact that Wittig recognized its significance is evidenced by the appearance of a second paper almost immediately coauthored with another research student, Ulrich Schöllkopf, titled "Triphenylphosphinemethylene as an Olefin-forming Reagent." In this paper it was shown that the reaction discovered accidentally was capable of being carried out under mild experimental conditions and could be used to prepare many "olefins" (a more modern term for olefins is "alkenes") that could not otherwise be obtained easily. The importance of the Wittig synthesis is that it allows the formation of olefins with the carbon-carbon double bond in a specific location in the molecule, which was not possible by any other procedure known in 1953 (or today, for that matter). Another advantage of the Wittig synthesis is that the mild conditions employed allow the production of substances that are sensitive to harsher conditions of temperature and stronger reagents.

Other papers came quickly from Wittig's laboratory, and also from laboratories of other chemists around the world. These research papers are too numerous to count, but a modern reference textbook lists two books and 25 review articles on the subject. Another criterion of the importance of the discovery, besides the Nobel Prize, is the fact that the Wittig synthesis has been used to prepare vitamin A by the ton!

Wittig's love of music exhibits itself in a paper he presented at an international symposium in Tokyo in 1964. He describes the history of the discovery in 1953 of the synthetic procedure that came to bear his name in a paper titled "Variations on a Theme by Staudinger." He said that the reaction he encountered accidentally was similar to one reported by Herman Staudinger 30 years earlier. He draws an analogy between his chemical "variation" on a "theme by Staudinger" and compositions of classical composers based on those of predecessors; for example, Brahms's "Variations on a Theme by Paganini." However, Wittig's discovery of a practical way to use organophosphorus compounds in synthesis was a revolutionary procedure, not a modification of an old process and, in fact, Wittig stated clearly that he was not aware of Staudinger's work until he searched the chemical literature after his serendipitous discovery.

The similar contributions of Brown and Wittig recognized by the Nobel Prize are their methods of using two noncarbon elements in synthesizing organic compounds of definite molecular structure by practical procedures. Both discoveries were the results of fortuitous accidents, followed by the brilliant extensions of these accidents in the prepared minds of Brown and Wittig.

▼▼▼▼▼▼▼▼▼▼▼▼▼▼▼▼▼▼▼▼▼▼▼▼▼▼▼

32 ▼

Polycarbonates: Tough Stuff

▼ Chance enters into chemical research in various ways to offer opportunities for discoveries. Sometimes it is the unknown presence of a contaminant in a starting material, as in the discovery of the crown ethers that led to Nobel Prizes for Charles J. Pedersen, Donald J. Cram, and Jean-Marie Lehn in 1987 (Chapter 36). Sometimes it is the accidental presence of a catalyst, as in the polymerization of ethylene by British chemists just in time to produce the ideal insulator for radar (Chapter 26). And sometimes it is as simple as not having the desired reagent in the storeroom and substituting something else for it.

That is what led to the discovery of a useful class of polyesters at a General Electric research lab. The polyester discovered is not the kind we wear, but still is one that we encounter almost every day. You might not know it by the name of polycarbonate, but you might have a bumper or a taillight on your automobile made of it, or you might sit on a jet airplane by a window made of it, or you might watch the president or the Pope ride in a vehicle protected by a shield made of it. When you encounter it next in one of these forms, be thankful that Daniel Fox did not find the substance he wanted in the storeroom that day in 1953.

Dr. Fox, who had completed his Ph.D. at the University of Oklahoma two years earlier, was given the job at the G.E. research laboratory in Schenectady of producing an insulating material for electrical magnet wire that would not decompose in high temperature and humidity. During one of the group planning sessions, someone rhetorically wished aloud, "If only we could find a hydrolytically stable polyester" (that is, one that is not decomposed by heat and humidity). This touched a memory chord in Fox; in his postdoctoral research project the preceding year he had found, to his surprise, that a carbonate ester of a phenolic compound known as guaiacol was unexpectedly resistant to hydrolysis. (Surprising, because most compounds of this type *are* decomposed by water.) So he went to the storeroom to obtain a chemical relative of quaiacol, bis-quaiacol, a compound that he thought would form links at both ends of its molecules and thus produce the desired polyester.

Fortunately, as it turned out, there was no bis-guaiacol in the storeroom, so Fox settled for a related compound called bisphenol-A. This material was cheap and readily available because it was an ingredient in epoxy resins, which had just come on the market. After several experiments in which alkyl carbonates were heated with bisphenol-A and failed to give anything promising, Fox heated an aromatic compound, diphenyl carbonate, with the bisphenol-A. As the expected by-product of the reaction, phenol (a high-boiling liquid), was distilled out of the reaction vessel, the contents of the vessel became more and more viscous. The temperature was raised and the pressure lowered progressively in order to remove more of the high-boiling phenol. This continued until the mechanical stirrer simply refused to turn any more as the motor driving it stalled.

Because the product was so thick that he could not pour it out of the flask, Fox allowed it to cool there. As it cooled, the product solidified and shrank away from the inside of the glass flask. Fox then broke the flask away from the solidified contents, leaving a hemispherical glob of solid on the end of a steel stirrer, with some adhering fragments of the glass flask embedded in it. In a 1987 speech, Dr. Fox described the next steps.

> This mallet [the glob of product] was pounded, and thrown on the cement floor without much effect. It was even used to drive some nails in a pine block. Eventually pieces were sawn off, [and] some were pressed into crude

▼ ***The first polycarbonate resin, formed around a stirrer in a glass flask that had to be broken away from it***

film at temperatures approaching 300°C. . . . Despite the fact there were a number of formally educated polymer chemists in the organization, most of the group had little-to-no actual background in synthesis of thermoplastic polymers, nor did they have any appreciation for the commercial significance of a polymer as different as this polycarbonate. The polycarbonates were shelved briefly. . . .

However, a Chemical Development Department, recently established in Pittsfield for the express purpose of developing new businesses was looking for projects. The manager of the development department at that time, Dr. A. Pechukas, had joined G.E. from Pittsburgh Plate Glass Co., where he had considerable aliphatic polycarbonate experience. Because of the contrast between PPG's (low-melting, brittle) aliphatic polycarbonates and G.E.'s (high-melting, very tough) aromatic polycarbonate, he became an instant champion for the early development of polycarbonates.

Developing the new plastic was not quick or simple, but eventually G.E. was successful and found numerous novel applications. The General Electric company was not the only one developing aromatic polycarbonates. The German company Bayer A. G. was also in the picture, although their process was slightly different. To be fair, Fox's discovery was actually a *re*discovery, because similar polymers had been prepared in Germany in 1902, although no use had been made of them and no one knew exactly what a polymer was in those days.

General Electric emphasized the clarity and toughness of the polycarbonates. Soon other plastics were billed as "almost as tough as polycarbonates" or "as tough as polycarbonates."

Polycarbonates are transparent, very tough, and capable of operation over a wide temperature range. Transparent bullet-resisting constructions are possible, including shields, that protect the president or Pope, foreign embassy structures, bank teller cages, barless windows in prisons, shields of priceless stained glass windows, spectator protection shields in hockey arenas, canopies for supersonic aircraft, scuba masks, and shields for SWAT teams.

Properties of strength, transparency, and steam-sterilizability enable these plastics to find applications from baby bottles to five-gallon water bottles, from medical equipment to goggles and helmets.

Laminates have been designed for Amtrak windshields capable of resisting an impact with a cinder block and for aircraft canopies capable of resisting bird impact at supersonic speeds. Less demanding applications include auto, bus, train, and airplane windows; luggage racks in airplanes; taillights and new design headlights for automobiles.

Blends with other polyesters have been formulated for crash-resistant bumpers on 35 different models of cars around the world.

Dr. Fox ended his 1987 speech with the statement, "It's been a lot of fun and a great source of personal satisfaction to see how far the glob on the end of the stirrer has gone and the lives which have been touched." I think we can all add our appreciation for his serendipitous failure to find the intended starting material, the fortuitous choice of its replacement, and the clever perception of Dr. Fox and his colleagues.

▼▼▼▼▼▼▼▼▼▼▼▼▼▼▼▼▼▼▼▼▼▼▼▼▼▼▼▼▼

33 ▼

VELCRO and Other Gifts of Serendipity to Modern Living

Velcro: From Cocklebурs to Spaceships

▼ The hook and loop fasteners known by the trade name Velcro are perhaps the world's most ingenious and versatile fastening method. And like many important discoveries, the idea for Velcro came from an accidental observation.

In the early 1950s George deMestral went for a walk in the countryside of his native Switzerland. Upon returning home he noticed that his jacket was covered with cockleburs. As he began picking them off, he wondered, "What makes them stick so tenaciously?" His curiosity led him to use a microscope to investigate more carefully. He discovered that cockleburs are covered with hooks, and the hooks had become embedded in the loops of the fabric of his cloth jacket. Nature's plan for dispersal and reproduction of the cocklebur plant is for the seed burs to become attached to passing birds and animals. DeMestral wondered whether a system patterned after the cocklebur could be designed that would be useful rather than a nuisance.

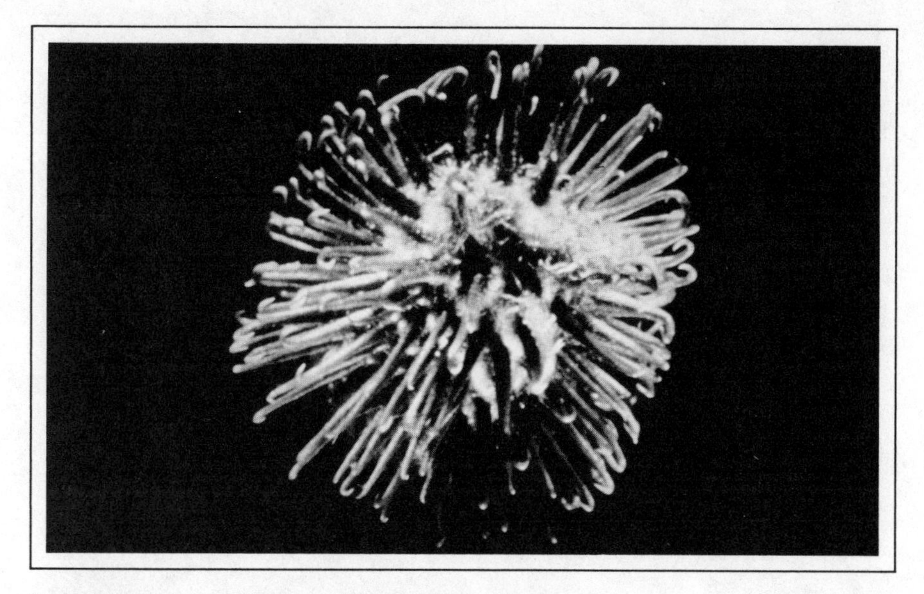

▼ ***A picture of a cocklebur enlarged to show the hooks***

The rest is fastener history. Today cocklebur-type hooks and woven loops secure everything from children's shoes to microphones in space shuttles. Velcro fasteners have been developed for applications in such diverse product areas as automobiles, home furnishings, medical supplies, military equipment, and many more too numerous to mention. The copyrighted name Velcro, by the way, is derived appropriately from *vel*vet and *cro*chet.

The first hook-and-loop tapes were made in France; production was by hand and agonizingly slow. Mechanical manufacturing of the loop tape was fairly easy, but making the hook tape was a formidable problem at first. The solution that made the system practical was to produce loops that could be cut near the ends to make a hook from each loop! Many improvements have been made since the original fasteners were put on the market; both the hook and the loop components have been upgraded by the use of modern materials. First the nylon filament used to make the hooks and loops was thickened; then the filament was made stronger by blending polyester with the nylon. Pure polyester tape was used to make it resistant to ultraviolet light, chemicals, and moisture. Even steel and space-age synthetic fabrics are now used to make fasteners that can be used in aircraft and space vehicles to withstand temperatures of 800°F. Some fasteners designed for space use are nonflammable even in pure oxygen.

▼ ***A photomicrograph showing the hooks and loops, the fundamental components of Velcro Brand fasteners***

Many of us have been annoyed by the cocklebu rs that stick to our clothing when we walk through the woods. We can all be thankful to George deMestral, who had the curiosity to find out how it happens and was ingenious enough to see a way to duplicate this feat of nature in such a way as to make the life of millions more convenient.

Ivory Soap

Among the myriad kinds of hand soaps, the only one that is unique is Ivory Soap, because "it floats!" The development of this unique soap was an accident. In 1879 an absentminded workman left a stirring machine running during the lunch hour and it whipped so much air into the batch of soap that the makers, Procter and Gamble, first considered throwing it out. But they hated to waste all that product, so they processed it and sold it. Much to their surprise, they began getting letters from a number of buyers asking for more of that miraculous floating soap.

Quick to recognize a sales gimmick, Harley Procter immediately began to advertise that the new floating soap could be used in the bath as well as the laundry. He notched the bars so they could easily be broken in half. He had the soap analyzed to compare it with three popular and expensive imported soaps and when the analysis showed P and G's soap

▼ *Floating soap*

contained the fewest impurities, he adversised it as 99 and 44/100% pure. And it has been that way ever since.

Oh, about the name: Harley Procter thought of that one Sunday in church as he read in Psalm 45 about "ivory palaces."

Corn and Wheat Flakes

The process for making both of these popular types of breakfast cereals was discovered somewhat by accident—enough to be called pseudoserendipitous at least. The corn flakes came first, in 1898. The Kellogg brothers (W.K. and J.H.) left some cooked wheat untended for more than a day, and when it was run through their rollers they were pleasantly surprised to find it came out flaked instead of in a flat sheet. The same procedure applied to corn gave them corn flakes, which became popular immediately. For some reason Wheaties did not appear for another 26 years.

Post-its

Divine inspiration, coupled with a product failure, gave birth to one of the 1980's top-selling and now indispensible products. People who use the usually yellow but possibly any-other-color self-sticking notes can't imagine what they did before without them. These ubiquitous slips of paper are found in offices on letters and file folders, by telephones and computer screens, in homes on the refrigerator and the TV screen, or possibly by the back door. These notes, which were invented at 3M and named Post-its, have encountered "the sincerest form of flattery," that is, they have been imitated and sold by everyone else.

But, back to the divine inspiration. In 1974 Art Fry was employed by the 3M company in product development. On Sundays he sang in the choir of North Presbyterian Church in North St. Paul, Minnesota. He marked his choir book in the time-honored way, with scraps of paper, to facilitate finding the proper music quickly at the proper time in the second service. But sometimes the scraps of paper fell out without warning, causing Fry to scramble belatedly through the pages in the second service.

"I don't know if it was a dull sermon or divine inspiration," says Fry, "but my mind began to wander and suddenly I thought of an adhesive, that had been discovered several years earlier by another 3M scientist, Dr. Spencer Silver." Spencer had discarded the adhesive, Fry remembered, because it was not strong enough to be permanently useful. Fry's inspiration was that this adhesive might serve to keep his place tem-

▼ ***A variety of Post-it notes***

porarily in the choir book without the marker becoming permanently attached—a "temporarily permanent adhesive," as Fry put it.

When Fry came to work on Monday and began making his bookmarks, it didn't take long before he began to envision other uses for them. He realized it was a system's approach to note making, because the notes had their own built-in means of attachment and removal. The idea was not an instant success, however. The adhesive had to be modified slightly to make it temporary enough and permanent enough, and this took quite a bit of experimentation.

After nearly a year and a half, Fry decided he had worked enough of the bugs out of the notes to show them to marketing personnel. Initially they were not overwhelmingly impressed; they were not sure that people needed a sticky note pad that would necessarily sell at a premium price compared to scratch paper. In 1977 Post-it notes were test marketed in four U.S. cities. In two of the cities the results were discouraging, but in the other two they were astonishing. Investigation as to the differing results disclosed that in the cities where reception was enthusiastic, dealers were handing out free samples. That was all it took—just getting the notes in the hands of the consumers. The rest is, to use a cliché, history. In 1980 Post-its were widely used throughout the United States, and by early 1981 sales in Europe paralleled those in the United States.

Scotchgard

According to 3M advertising, their Scotchgard brand of fluorochemical stain repellant was an accidental discovery. Someone spilled a bit of the newly developed product on a tennis shoe, noticed that the area of the spill could not be easily soiled, and had the sense to think what that might mean. Another accident became a discovery.

▼▼▼▼▼▼▼▼▼▼▼▼▼▼▼▼▼▼▼▼▼▼▼

34 ▼

DNA: The Coil of Life

▼ Most people have a vague idea that the letters DNA are somehow important, and some even know that they are the acronym of *d*eoxyribo*n*ucleic *a*cid, but few nonscientists know the significance of the substance bearing this jaw-breaking name. The knowledge of its molecular structure and biological function has been called the secret of life.

In 1962 James Watson, Francis Crick, and Maurice Wilkins shared the Nobel Prize in Physiology or Medicine for their research that revealed the structure of DNA. Watson was a biologist, Crick a physicist, and Wilkins an X-ray crystallographer. The details of their work are described in every textbook of organic chemistry and biochemistry; however, none of these books mentions that an event of serendipity involving a chemist was a key to the breakthrough that led to their Nobel Prize. Watson himself disclosed the serendipity in his exciting autobiographical account of the discovery, *The Double Helix.* This title graphically describes the structure of this vital molecule.

James Watson received his Ph.D. in biology at Indiana University in 1950 when he was only 22. He went to Europe for postdoctoral study and research, and in the second year he was at Cambridge University in the laboratory of Sir Lawrence Bragg, a Nobel laureate in 1915 for his use of X rays in determining the structure of crystals. At Cambridge, Watson spent most of his time in collaboration with Francis Crick, a brilliant and somewhat unorthodox physicist. These two decided to pool their interdisciplinary assets with the frank intent of winning a Nobel Prize by solving the mystery of DNA.

They took as their guide the approach of Linus Pauling to discovering the α-helix structure of proteins (a linear right-handed helical or spiral alignment of the atoms in the giant molecules). Pauling received the Nobel Prize in 1954 for his work, and Watson and Crick could see this coming even in 1952. Pauling's approach was to rely on laws of structural chemistry, many of which he developed himself, and to apply these to models of proteins that superficially resembled toys of preschool children. Pauling's models, however, were designed to fit sizes and shapes based on X-ray pictures of crystals.

DNA was known to be a giant molecule like a protein, and the X-ray data on it were not incompatible with a helical, or coil, structure. The best X-ray data available to Watson and Crick came from the laboratory of Maurice Wilkins at Kings College, a part of the University of London, and was produced by the collaboration of Rosalind Franklin with Wilkins. Protein molecules are made up of many amino acid units (monomers, from the Greek for "single parts") joined together to make the giant molecule, or polymer (Greek for "many parts"). Analyses of DNA had shown by 1952 that it was a polymer made of more than one kind of monomer. The repeating monomer units were deoxyribose (a kind of sugar), phosphoric acid, and four different organic bases—guanine, cytosine, adenine, and thymine.

One of the clues that helped Watson and Crick solve the puzzle of DNA was provided by an Austrian-born chemist at Columbia University, Erwin Chargaff. Chargaff reported that in his studies of DNA from various living sources he always found a one-to-one relationship between guanine and cytosine and between adenine and thymine. In other words, guanine was always paired with cytosine and adenine was always paired with thymine.

Watson and Crick developed a model of DNA in which there were two helices, or coils, made up of deoxyribose and phosphoric acid units on the outside that were held together in some way by the organic bases on the inside of the double spiral. They planned to use models of the component units made in the machine shop of the Cavendish Laboratory of Cambridge University, and to put the pieces of the model together in such a way as to satisfy the measurements obtained from X-ray pictures of DNA crystals. While waiting for the machinists to make the metal model units, Watson busied himself with drawings of the bases and making cardboard models. He came to the conclusion that the bases, which were known to be regularly repeating components of the outside coils of the polymeric molecule, somehow held the coils together by hydrogen bonds between pairs of bases in a like-with-like relationship. ("like-with-like" means guanine-with-guanine, cytosine-with-cytosine, and so on).

Hydrogen bonds are a type of chemical bond, or connection, between atoms that are known to be capable of holding together molecules such as the bases present in DNA. Water molecules are also held together in groups by hydrogen bonding; therefore, the effective molecular size of water is much larger and the volatility much lower than if water existed as single molecules of H_2O. If this were not so, we would live in a very different world—there would be no liquid water on our planet!

Watson knew, however, that the DNA bases can exist in different tautomeric forms, which means that a hydrogen atom responsible for making a bond can be in more than one position in the molecule of the base. Watson used formulas taken from a current book that he assumed placed the hydrogen atoms of the bases in the correct positions, and he joined the parts of the coils together with hydrogen bonds between the bases in a like-with-like manner. The model produced in this way appealed to Watson because he saw in it a way of explaining the known process of replication by genes, but it failed to satisfy the molecular dimensions dictated by X-ray data or the Chargaff rule of pairing of guanine with cytosine and adenine with thymine. Although slightly dissatisfied with the proposed structure, because of the urgency of beating Linus Pauling and perhaps others to the Nobel Prize, Watson posted a letter to a colleague in which he claimed that he had "just devised a beautiful DNA structure."

No more than an hour after posting this letter, Watson encountered Jerry Donohue, an American physical chemist and crystallographer, in his office and began explaining his theory to him. Donohue happened to share the office with Watson and Crick at that time. Donohue immediately threw cold water on Watson's structure: he said that Watson had used the wrong tautomeric forms of the bases for the hydrogen bonding that held the spiral chains together. Watson protested that not only the book he referred to, but others as well, pictured the bases in the tautomeric forms he had used. Donohue said that wrong tautomeric forms had been published for years with little real evidence to support them. Because Donohue had done X-ray crystal studies of molecules like the DNA bases at Cal Tech (Pauling's home base), he spoke with authority when he told Watson he had used the wrong tautomeric formulas.

Watson went back to his desk and made new cardboard models based on the other (correct, according to Donohue) tautomeric formulas. He found that they did not work at all! When he showed Crick the structure as modified according to Donohue's corrections, Crick also realized that it was unsatisfactory in fitting the X-ray dimensions and would predict violations of Chargaff's rule.

A thoroughly discouraged Watson went home that night, but he was back at his desk early the next morning. In *The Double Helix* he describes what happened then:

> When I got to our still empty office the following morning, I quickly cleared away the papers from my desk top so that I would have a large, flat surface on which to form pairs of bases held together by hydrogen bonds. Though I initially went back to my like-with-like prejudices, I saw all too well that they led nowhere. When Jerry came in I looked up, saw that it was not Francis, and began shifting the bases in and out of various other pairing possibilities. Suddenly I became aware that an adenine-thymine [A-T] pair held together by two hydrogen bonds was identical in shape to a guanine-cytosine [G-C] pair held together by at least two hydrogen bonds. All the hydrogen bonds seemed to form naturally; no fudging was required to make the two types of base pairs identical in shape. Quickly I called Jerry over to ask him whether this time he had any objection to my new base pairs. When he said no, my morale skyrocketed, for I suspected that we now had the answer to the riddle.

He later added: "Our idea was aesthetically elegant. . . . a structure this pretty just had to exist." Figure 34.1 shows the base pairs Watson was working with.

Watson and Crick quickly submitted a short paper to *Nature.* It began with the modest statement: "We wish to suggest a structure for the salt of deoxyribose nucleic acid (DNA). This structure has novel features which are of considerable biological interest." The second sentence must qualify as one of the best examples of understatement in print.

Toward the end of his book, Watson wrote: "The unforeseen dividend of having Jerry share an office with Francis, Peter (Pauling), and me, though obvious to all, was not spoken about. If he had not been with

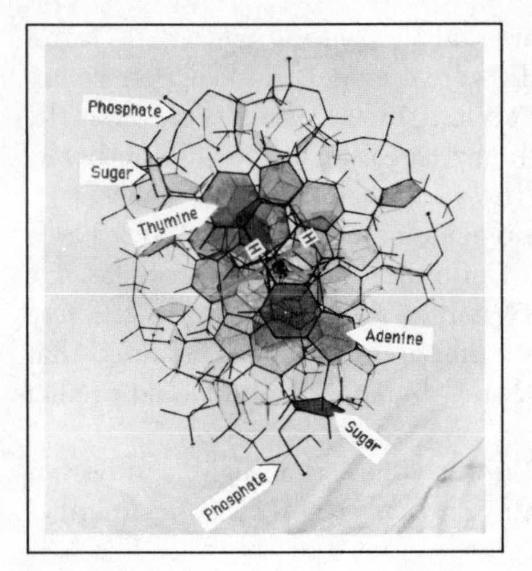

▼ *Top view of a model of a DNA molecule. Compare with Figure 34.1.*

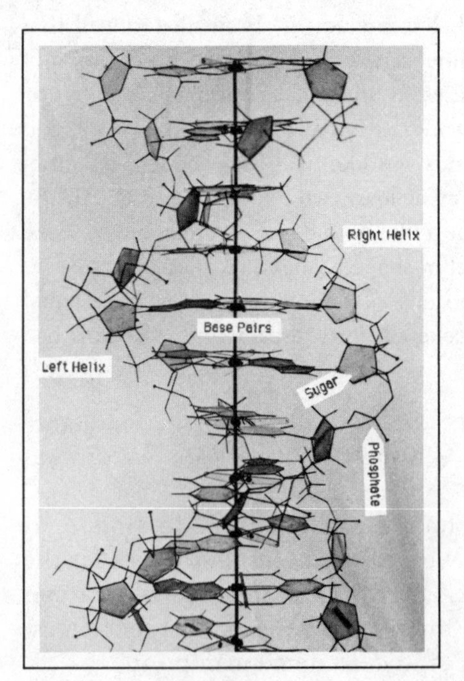

▼ *Side view of the model of a DNA molecule*

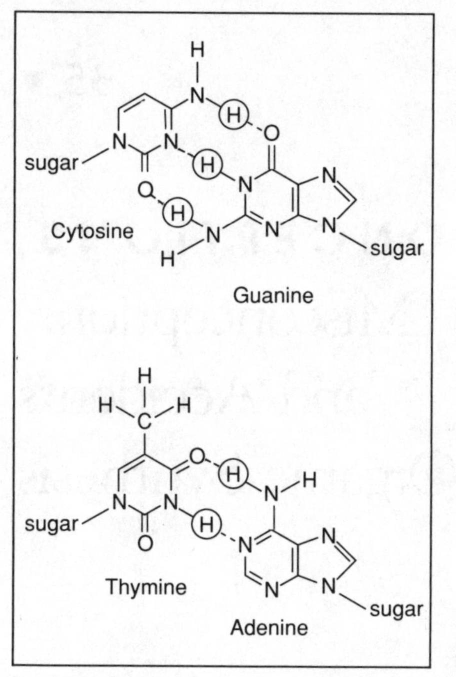

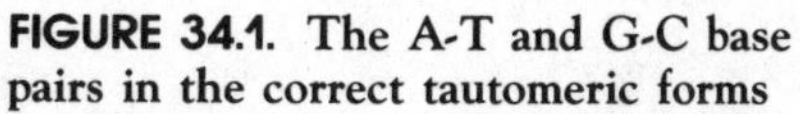
FIGURE 34.1. The A-T and G-C base pairs in the correct tautomeric forms

us in Cambridge, I might still have been pumping for a like-with-like structure."

So the serendipity in the DNA story is the "dividend of having Jerry share an office with Francis . . . and me." Another example of the importance of serendipity to a Nobel Prize-winning discovery!

35 ▼

Conceptions, Misconceptions, and Accidents in Organic Synthesis

▼ Sir Derek H. R. Barton shared the Nobel Prize for Chemistry with Odd Hassel of Norway in 1969. Barton, an organic chemist at London's Imperial College of Science and Technology, and Hassel, a physical chemist retired from a professorship at Oslo University, were honored for their work in *conformational* analysis, the study of the three-dimensional relationships of molecules. Conformation has sometimes been referred to as the "fourth C-word" describing the structure of organic molecules, the other three being *composition, constitution,* and *configuration.* The last two of these terms describe the three-dimensional aspects of molecular structure. (For more information about the configuration of organic molecules, see Chapter 12, on Pasteur's discovery of right- and left-handed molecules.)

Odd Hassel was born in Oslo in 1897. He was educated at the University of Oslo in 1915–1920. In 1922–1924 he worked with Professor Herman Mark at the University of Berlin, where he received his Ph.D., and then he returned to the University of Oslo. Although he

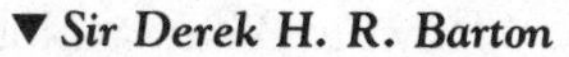
▼ *Sir Derek H. R. Barton*

wanted to continue research in X-ray crystallography, which he had learned with Mark, he had to be content in measuring dipole moments at Oslo because the X-ray equipment was not available at first. The dipole moment studies could be done with the simpler instruments Hassel had access to. This may have been fortuitous, because this dipole work, which began with measurements on six-carbon ring compounds (cyclohexanes), led directly to the work that was recognized by the Nobel Prize.

From the time of Adolph von Baeyer in 1885 until the 1920s, most chemists considered the six-carbon ring of cyclohexane to be planar, as shown in Figure 35.1a. However, between 1918 and 1925 two different puckered nonplanar shapes were established as correct. These were called boat (Figure 35.1b) and chair (Figure 35.1c) conformations, the latter being preferred because of certain repulsions between the atoms in the

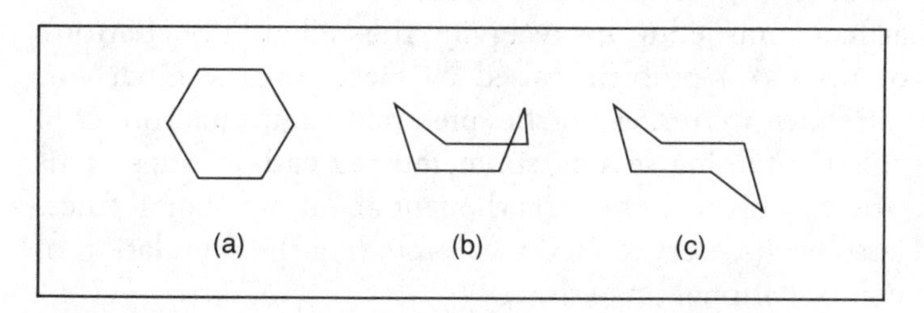

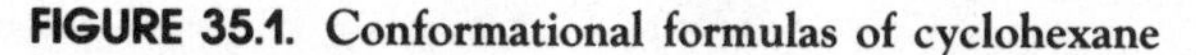
FIGURE 35.1. Conformational formulas of cyclohexane

boat form. By 1943, Hassel had drawn the conclusion from his results that virtually all compounds containing the six-membered ring of cyclohexane had the chair shape, but he published his conclusion in Norwegian in an obscure Norwegian research journal where it was generally overlooked.

Shortly after this, the Germans invaded Norway and Hassel was imprisoned by the Nazis for two years in a concentration camp. After the war he returned to the University of Oslo as a professor and remained there until he retired in 1964, retaining a connection with the university, however, until his death in 1981. By an interesting coincidence, while in prison Hassel was a cellmate of Ragnar Frisch, who also received a Nobel Prize (for economic science) in 1969, the same year Hassel shared the prize with Barton for chemistry.

At a conference in Germany, Barton gave an interesting history of his and Hassel's chemical contributions. The title of his paper is self-explanatory: "How to Win a Nobel Prize: An Account of the Early History of Conformational Analysis." Some of the preceding material and that following about Hassel and himself comes from that article.

Derek Harold Richard Barton was born in 1918 in Gravesend, England. Both his father and his grandfather were carpenters. His father founded a wood business that was successful enough to allow his son to go to a good school (Tonbridge), but he was forced to leave it "without any qualifications" upon the sudden death of his father in 1935. Two years in the wood business convinced young Derek that there were more interesting things in life, and within a few years he had a B.Sc. and a Ph.D. in organic chemistry from the Imperial College of Science and Technology, a part of the University of London. Two years in military intelligence followed during the war, and then a year in industry, after which he jumped at the opportunity to return to Imperial College to teach.

In the spring of 1949 he was invited by Professor Louis Fieser to come to Harvard to take a position temporarily vacated by R. B. Woodward, who was on sabbatical. During the year at Harvard, Barton wrote a revolutionary paper (short, according to Barton, because he had to type it himself!) that he claims led to his receiving the Nobel Prize. Barton's paper was a solution to a problem raised by Fieser in a seminar—to analyze Fieser's results in terms of the preferred conformation of a steroidal molecule containing several six-membered carbon rings. This paper changed the way organic chemists thought about many molecules. Barton gives Hassel well-deserved credit for providing the foundation of this theory of conformational analysis.

Barton returned to England for positions at the Birbeck College of the University of London, at Glascow University, and then back to Imperial College in 1957, where he held the Chair of Organic Chemistry for 22 years. In 1978 Sir Derek resigned to become director of the Institute for the Chemistry of Natural Products, C.N.R.S., in a suburb of Paris. He left that Institute in 1987 to take a Chair at Texas A&M University at College Station, Texas.

Sir Derek has categorized his research in terms of processes of conception, misconception, and accident. The misconceptions and accidents he cited are perfect examples of serendipity, and the conceptions described illustrate clearly the importance of planned (*conceived*) research. Although my emphasis in this book is on the accidents that have led to valuable discoveries, I have tried to show in each case that the accident would not have led to a discovery but for the "prepared mind" (Pasteur) and "sagacity" (Walpole) of the person who encountered the accident.

As an example of a conceived synthetic process, Sir Derek describes the conversion of the readily available compound, corticosterone acetate, into aldosterone, a hormone largely responsible for maintaining the normal balance between salts in the human body. At the time of the conception of the photochemical process by Barton, the world supply of pure aldosterone was measured in milligrams (a small drop of water weighs about 50 milligrams). One year later Barton and his co-workers had 60 grams (60,000 milligrams) of the pure hormone, allowing proper evaluation of its biological activity for the first time.

A process that turned out to be useful in the synthesis of other naturally occurring substances was discovered by misconception. The original procedure was designed to produce in a complex molecule a cyclic unit called a γ-lactone. However, it gave no γ-lactone, but an entirely unexpected result. After careful analysis of the actual course of the reaction, Barton and his co-workers were able to use this new reaction for other valuable syntheses. Frustrated still by not having achieved the original goal, however, Barton decided to try again to make the γ-lactone by modifying the starting materials and reactants. He was successful this time, and a few years later in collaboration with Jack Baldwin (who is now a professor at Oxford), he used the modified procedure to convert estrone (Figure 35.2a) into 18-hydroxyestrone (Figure 35.2b), "a compound found in such minute amounts in female urine that its biological activity (like that of aldosterone) could only be evaluated properly after being made available by synthesis." Thus Sir Derek's original mis-

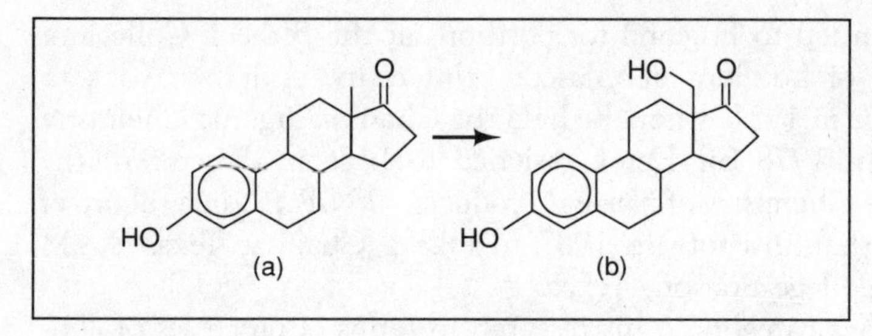

FIGURE 35.2. Molecular formulas of estrone and 18-hydroxyestrone

conception led to two new and important conceptions that were valuable in biorganic chemistry.

This paper mentions several examples of accidents, and Sir Derek described another in a seminar at the University of Texas at Austin in the fall of 1984. (Sir Derek began this seminar with the comment, "You know, most of the important reactions in organic chemistry were discovered accidentally," and he proceeded to give several examples before describing his own latest encounter with serendipity. At that time, I had been interested in serendipity and its contributions to science for many years, and soon after that seminar I began to think seriously about putting together a book on the subject.) In the previous year, he and his coworkers at the Institute for the Chemistry of Natural Products decided to prepare some model compounds to compare with a complex natural product they were studying. One of the model compounds desired had the formula shown as (b) in Figure 35.3. It was expected to be the

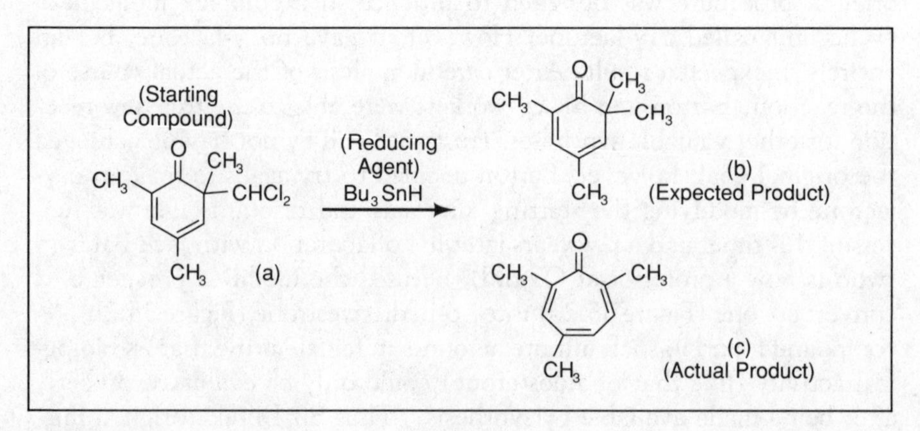

FIGURE 35.3. Accidental synthesis of a tropone

product formed by reaction of compound (a) in Figure 35.3 with the reducing agent tributyltin hydride—two hydrogen atoms would replace the two chlorine atoms in (a) converting the $CHCl_2$ group into a CH_3 group. To Barton's utter surprise, not even a trace of the expected product (Figure 35.3b) was obtained, but instead a nearly quantitative yield of a compound in which the six-membered carbon ring had expanded to a seven-membered ring (Figure 35.3c).

They recognized the actual product as a member of a group known as *tropones,* which have been of great interest to organic chemists and biochemists. Tropones and tropolones (closely related compounds) are intriguing partly from a theoretical viewpoint, having the unusual seven-membered ring as a part of their molecular structure, and partly because they are naturally occurring substances that have valuable physiological properties. Colchicine, for example, has been used effectively in treating gout and, because it increases the number of chromosomes during cell division in plants, it has been used commercially to produce new plant species. Tropones and tropolones have not previously been easy to synthesize. Because the starting compound (a) in Figure 35.3 and similar compounds are readily prepared from simple starting materials, Barton and his co-workers recognized this accidental discovery as a generally useful synthetic procedure.

Identifying the tropone as the unexpected product of the reaction done for another reason was not a trivial achievement, and explaining the mechanism of the reaction was a superb exercise in the logic of organic chemistry. The unexpected result is probably more important than the conceived result of the reaction would have been; therefore, this is a true example of serendipity.

36 ▼

CHEMICAL Crowns and Crypts

▼ The Nobel Prize for Chemistry in 1987 was shared by three men, two Americans and a Frenchman. These men represent three different generations of chemists: Charles J. Pedersen received a master's degree in chemistry from M.I.T. in 1927; Donald J. Cram received a Ph.D. in chemistry from Harvard in 1947; Jean-Marie Lehn received a doctorate in chemistry from the University of Strasbourg in 1963. Pedersen made the basic discovery by accident in the 1960s and the other two men later extended his work to imaginative bioorganic applications. Pedersen is proud that his prize-winning research was done late in his career at Du Pont. After learning of his award he said, "It tends to be said that the best work performed by scientists is done by the time they are 35 years old. This work was done in my last nine years at Du Pont." Pedersen was 63 when his research was published. He retired two years later.

Pedersen's discovery came about because a contaminant was accidentally present in one of the chemicals he used in an experiment. Pedersen turned this accident into a discovery of tremendous importance by observing the unexpected appearance of "a white, fibrous, crystalline by-product," isolating and examining it, and recognizing its unusual properties. He found that this material could combine with inorganic salts like sodium chloride and potassium chloride and make them soluble in organic liquids in a way that had not been possible before.

Having found the by-product accidentally, Pedersen was able to make it intentionally, and he proceeded to make a whole series of other compounds like it. These new compounds were cyclic ethers, related structurally to diethyl ether, the well-known anesthetic. However, instead of having one oxygen atom per molecule, they have several oxygen atoms separated from one another by two or more carbon atoms in a many-membered ring. When combined with an inorganic salt, these cyclic ethers form a complex in which the cyclic ether is held around the metal part of the salt like a crown resting on a human head. Therefore, Pedersen called the new cyclic ethers "crown ethers."

Figure 36.1 shows the formula of the first crown ether discovered by Pedersen "crowning" a metal ion, M^+ (the positive part of the salt, which might be the Na^+ or K^+ ion). In the figure only the oxygen atoms and the metal cation are indicated by letters; each corner in the drawing represents a carbon atom with one or two hydrogen atoms attached. The figure is based on a three-dimensional X-ray picture of a crystal of the complex. Figure 36.1a is an overhead view. Figure 36.1b is a perspective view from in front of and slightly above the complex; it shows more clearly the resemblance of the molecule to a crown.

The size of the ring, the size of the metal ions, and the number of oxygen atoms in the polyether ring determine the stability of the complexes.

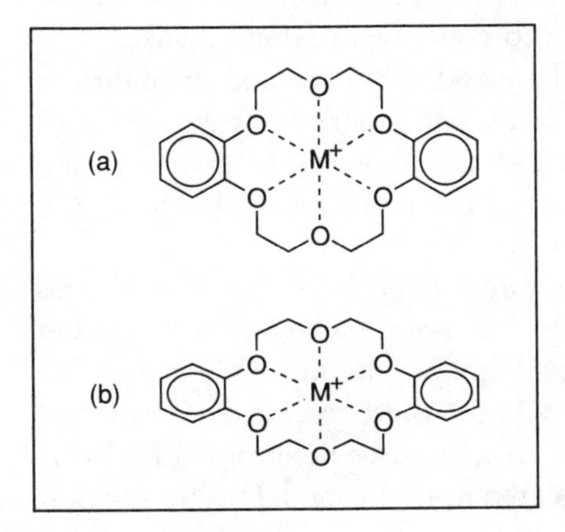

FIGURE 36.1. Molecular formulas of a dibenzo-18-crown-6 complex

Although the formation of the first crown ether was accidental, Pedersen had the "prepared mind" necessary to turn this accident into a major discovery. In his research at Du Pont he had for many years investigated the ability of certain organic molecules to combine (chelate) with metals, and had used chelation to remove small amounts of metals that contaminate gasoline and other petroleum products. He held many patents covering antioxidants, stabilizers, and inhibitors based on the chelation of metals with organic compounds. Thus, it was natural for him immediately to turn his attention to the metal-complexing properties of the new crown ether he had produced, and to synthesize analogous crown ethers and compare them with the first one.

Reaction to Pedersen's first published description of the crown ethers (in 1967) was fast and enthusiastic, worldwide. Jean-Marie Lehn in France extended Pedersen's complexes to three dimensions and incorporated elements besides oxygen (such as nitrogen) in the rings. The three-dimensional shape of Lehn's molecules made them more rigid and increased the range of substrates that they could bind. He invented new names for his complexes, calling the three-dimensional complexes *cryptates*, and the host components *cryptands;* he derived these names from the Greek word *kryptos* for "hidden."

Donald J. Cram at U.C.L.A. extended Pedersen's work to establish a new field that he called "host-guest" chemistry, in which the host is the receptor molecule and the guest is the substrate received. Although Lehn and Cram used different approaches, they both applied a concept of synthesizing organic molecules that were sufficiently rigid and contained cavities of a size and shape so as to accommodate certain desired substrates. For example, they envisioned preparing comparatively simple catalysts (in contrast to the enormous molecules of biological enzymes) that would mimic the actions that enzymes perform in cells, functions such as splitting protein molecules into constituent amino acids.

Because host-guest chemistry is based on the shape and flexibility or rigidity of three-dimensional molecules, Cram has made extensive use of molecular models, such as are shown in the photographs on the facing page. These photographs may be compared to the formulas in Figure 36.1.

These models are made of plastic interlocking balls; the relative sizes of the balls and the angles at which they are attached to one another represent rather accurately the actual known atomic dimensions and bond angles. Having prepared some 90 host-guest complexes, Cram and his colleagues have found that their structures, as determined by X-ray diffraction, are approximately as expected from the models in most cases.

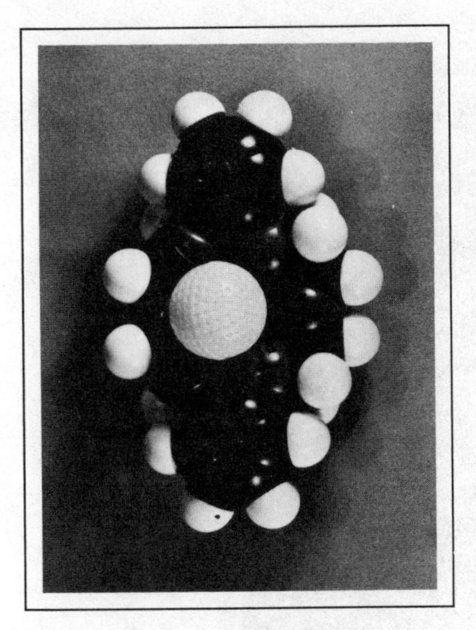

▼ ***A space-filling model of the dibenzo-18-crown-6 complex, top view***

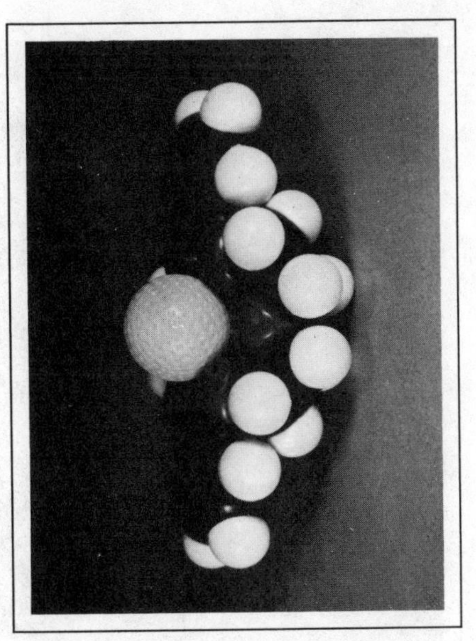

▼ ***A space-filling model of the dibenzo-18-crown-6 complex, side view***

This field of molecular recognition in organic chemistry is growing in both basic and applied research areas. It has yielded significant insights into chemical reactions that occur within human cells, and has laid the foundation for a field of biomedical research that has grown rapidly. It was the topic of a symposium at the national meeting of the American Chemical Society in Denver, Colorado, in spring 1987, where chemists from several universities and pharmaceutical companies described their latest work, all of which could be traced back through Cram and Lehn's research to Pedersen's discovery.

When Pedersen was awarded the Nobel Prize, which he shared with Cram and Lehn in November 1987, he was 83 and in poor health. After returning from Stockholm, where he participated in the Nobel ceremonies, he was interviewed in his home by Shekhar Hattangadi for *Industrial Chemist;* the following quotations are from that interview:

> Imagine this sequence of events, ca. 1900: An engineer in Norway decides to go halfway around the world to Korea, where he works in a gold mine. A Japanese family, having suffered some financial reverses in Japan, decides to move to Korea, where markets are opening up. The brother starts a business close to the mine. The sister meets the young Norwegian; they marry. Some

▼ *Charles J. Pedersen*

years later, their son travels to the United States for his education. He becomes a chemist and wins the Nobel Prize.

"It's the most unlikely scenario I can think of," exclaims Charles Pedersen. . . , recounting his family history. "I spoke about this in Stockholm."

He does not recall . . . the precise details of the methodology that led to his historic discovery of the crown ethers. "The discovery was simply the result of this: I did the right thing at the right time," he says. "But I cannot tell you right now exactly why I did what I did." . . .

Over the years, Pedersen has lived by two . . . truths: First you don't have to be a Ph.D. to do good research. Second, you don't have to be young to do your best work. "If an old man can get ideas, you should let him try them," he says, mindful that he began research on crown compounds in his 60s.

> Herman Schroeder first met Charles Pedersen 48 years ago, as a colleague in Du Pont. They have been friends ever since. At a ceremony in Japan to honor Pedersen on the 20th anniversary of his discovery of crown ether compounds, Schroeder filled in for the ailing honoree and made a moving speech outlining his friend's achievements. . . . "You should understand the crown ether discovery in perspective. Charlie was an acknowledged expert in coordination chemistry. . . . He knew what could or couldn't happen; he was, in a sense, prepared for the discovery. It wasn't that he stumbled on it; rather, it was as if the compound walked in front of him and he snapped it up. You need a sharp, ready, and flexible mind for that."

It is another comfirmation of what I have noted so many times in these accounts of accidental discoveries—an accident is an accident until it happens to the right person—then it can become a discovery!

EPILOGUE: How Accidents Become Discoveries

The subtitle of this book is "Accidental Discoveries in Science." You have probably realized that, in every example of accidental discovery in this book, accidents became discoveries because of the sagacity (Walpole's word) of the person who encountered the accident. In this epilogue I explain what is meant by the sagacity needed for discovery and suggest how it may be developed and encouraged.

Pasteur said, "In the fields of observation, chance favors only the prepared mind." What is a "prepared mind," and how can it be acquired? I feel sure that there is an inborn ability or talent for discovery in many of those who benefit from serendipity, persons like Priestley, Pasteur, and Perkin. The dominant characteristic shared by these three, and by many others who turned accidents into discoveries, is curiosity. They were curious to understand the accident they had observed.

Another characteristic they shared is perception. They observed a phenomenon that was unexpected, and they took note of it rather than dismiss it as trivial or annoying. Undoubtedly many persons had seen a gas produced unexpectedly, crystals with an unusual shape, a deep color in a waste material being discarded—but they did not discover oxygen, the most fundamental element, did not deduce a type of molecular structure vital to life itself, or develop a beautiful dye from black coal tar. Albert Szent-Gyorgyi said it well: "Discovery consists of seeing what everybody has seen and thinking what nobody has thought."

Although curiosity and perception may be more inherent in some persons than in others, they can be encouraged and developed, Ronald S. Lenox has suggested. In an article titled "Educating for the Serendipitous Discovery" (*Journal of Chemical Education,* Vol. 62 1985, p. 282), Lenox describes several ways in which students can be prepared to take advantage of fortuitous accidents. The first way is to provide training in making and recording observations, including *unexpected* as well as expected results. Lenox suggests that such training should require the student to keep a laboratory notebook that the instructor will grade on the basis of not only "correct" or "incorrect" answers, but also on the basis of observational and recording skills.

Students should be encouraged to be flexible in their thinking and interpretations. The person who sees only what is expected and discards unexpected results as "wrong" will make no discoveries. In my own experience I have had two good examples that illustrate this principle. One graduate student carried out an experiment and reported a result I expected. Later, a second student could not reproduce this result, and it was subsequently found to be incorrect. The first student was under pressure of time and reported, incorrectly, the result his professor had told him to expect. (Fortunately we did not publish the incorrect data, but waited until later when the problem had been cleared up.) In a second case, a student was told he might expect a certain result, but he observed something he recognized to be quite different, and he reported correctly what he had observed. The unexpected result led to a whole new concept of the originally planned research program—to a *discovery.*

Another way in which a person can prepare to benefit from serendipity is through careful and intensive study in the field of chosen investigation. The American physicist Joseph Henry paraphrased Pasteur's dictum when he said, "The seeds of great discovery are constantly floating around us, but they only take root in minds well prepared to receive them." For example, although Fleming was not looking for an antibacterial agent at the time a spore floated into his petri dish, he was extremely well read and trained in microbiology and could readily recognize the meaning of the clear area in the bacterial culture produced by the accidental implantation of the mold.

Some of the discoveries I have characterized as pseudoserendipitous would not have become discoveries if the investigator had not been prepared to recognize an accident as one that yielded the hoped-for result, although by a totally unexpected and accidental occurrence. For example, Goodyear quickly recognized that the accident of heating rubber with sulfur had the effect of making the elasticity of rubber remain unaffected by heat and cold, an effect he had been seeking for many years, an effect he knew how to test for immediately. Similarly, Nobel saw in the accidental spill of nitroglycerine into the porous packing material around it a possible solution to the dangerous nature of the explosive that he had sought so desperately. His mind made the leap to discovery because he had experimented with mixtures of the capricious liquid with other materials before and knew just how to test this mixture formed by accident.

In contrast to a few such pseudoserendipitous discoveries are the truly serendipitous discoveries: fortuitous accidents that led to new things totally unexpected and unsought for, and which became *discoveries* through the sagacity of the person who encountered the accident. One cannot deny the importance of "being in the right place at the right time," but one should note carefully that in Walpole's explanation of serendipity, he gave equal billing to "accidents" and "sagacity." He told his friend Sir Horace Mann that the three princes of Serendip "were always making discoveries, by accidents and sagacity, of things which they were not in quest of."

Consider some of the accidents I have described: an apple falls to the ground at Newton's feet; Wöhler produces urea instead of ammonium cyanate; Perkin makes a violet color; an Italian well-digger finds statuary instead of water; a French soldier turns up a stone with strange inscriptions while repairing a fort in Egypt; a Palestinian boy throws a rock into a dark cave and hears an unexpected sound; Pasteur obtains crystals of unique shape because the temperature on the window ledge was below 79°; a fluorescent screen glows in the dark when it shouldn't and phosphorescent crystals expose a photographic plate wrapped in black paper; a milkmaid's resistance to a dread disease is noticed by a country doctor; a spore falls into Fleming's open petri dish; Du Pont chemists play games in the laboratory hall and find that a stretched fiber is unusually strong; another Du Pont chemist finds that no gas comes out of a tank even though the tank is not empty; leaky and unclean equipment leads to the formation of polymers when none was expected; the accidental presence of an impurity in a starting material produces a novel molecular "crown."

Any of these accidents could have gone unnoticed and would have thus remained simply an accident of no importance. Instead, because of the sagacity of the individuals who encountered the accidents, we have an explanation of the laws that govern the movement of the planets; the founding of the science of organic chemistry on a rational basis; the beginning of understanding the relationship of molecular structure to physiological activity; beautiful dyes that anyone, not just royalty, can afford; insight into the culture and language of ancient civilizations; X rays for medical diagnosis and treatment; radioactivity and nuclear energy; vaccination against smallpox and other diseases; the "miracle drug" penicillin and its successors; nylon and polyester for clothing; Teflon for frying pans and heart valves; other polymers for plastic garbage bags, ice chests, radar insulation, water-ski ropes, bullet-proof shields, and airplane windows; and synthetic molecules that promise to mimic the vital actions of nature's enzymes.

These are just a few of the benefits of serendipity, "discoveries made by accident and sagacity of things which they [certain gifted individuals] were not actually in quest of." Some of these discoveries were made centuries ago, some recently. In the twentieth century our knowledge in science, medicine, and technology has grown at a fantastic rate. We cannot conceive what advances the future may bring—interplanetary space travel? a cure for cancer? But we can be sure that accidents will continue to happen and, with human minds better prepared than ever before, we can expect these *accidents* to be turned into *discoveries*, marvelous beyond our imagination, through serendipity.

Appendix: References and Additional Reading

1. ARCHIMEDES—THE FIRST STREAKER

"Archimedes." *Encyclopaedia Britannica.* IX Edition. 1878. Vol. 2.

"Archimedes." *Encyclopaedia Britannica.* 1962. Vol. 2.

2. COLUMBUS DISCOVERS A NEW WORLD

"Columbus, Christopher." *Encyclopaedia Britannica.* 1962. Vol. 6.

Judge, Joseph and J. L. Stanfield; Luis Marden; Eugene Lyon. "Columbus and the New World." *National Geographic,* November 1986.

3. A SICK INDIAN DISCOVERS QUININE

Encyclopaedia Britannica. 1962. See "Cinchona" (Vol. 5) and "Quinine" (Vol. 18).

Accidental Scientific Discoveries. Chicago: Schaar and Co., 1955. P. 45.

Lednicer, Daniel, and Lester A. Mitscher. *The Organic Chemistry of Drug Synthesis.* Vol. 1. New York: Wiley-Interscience, P. 337.

Silverman, M. *Magic in a Bottle.* New York: Macmillan, 1941. Chapter 2.

4. SIR ISAAC NEWTON, THE APPLE, AND THE LAW OF GRAVITATION

Andrade, E. N. da C. "Newton." In *Newton Tercentenary Celebrations.* The Royal Society. Cambridge: Cambridge University Press, 1947.

Andrade, E. N. da C. *Isaac Newton.* London: Max Parrish, 1950.

Anthony, H. D. *Sir Isaac Newton.* London: Abelard-Schuman, 1960.

Brewster, Sir David. *Memoirs of the Life, Writings, and Discoveries of Sir Isaac Newton.* Vols. I and II. Edinburgh: Edmonston and Douglas, 1850.

Cajori, Florian. "Newton's Twenty Years' Delay in Announcing the Law of Gravitation." In *Sir Isaac Newton, 1727–1927, A Bicentenary Evaluation of His Work.* The History of Science Society. Baltimore: Williams and Wilkins, 1928.

de Villamil, R. *Newton: The Man.* London: Gordon D. Knox, 1931.

More, L. T. *Isaac Newton: A Biography.* New York: Charles Scribner's Sons, 1934.

Stukeley, William. *Memoirs of Sir Isaac Newton's Life.* 1752. Reprint. London: Taylor and Francis, 1936.

Sullivan, J. W. N. *Isaac Newton, 1642–1727.* New York: Macmillan, 1938.

5. THE ELECTRIC BATTERY AND ELECTROMAGNETISM—FROM A FROG'S LEG AND A COMPASS

Encyclopaedia Britannica. 1962. See "Battery" (Vol. 3), "Electromagnet" (Vol. 8), and "Oersted" (Vol. 16).

Bailar, John C., et al. *Electrochemistry.* New York: Academic Press, 1978. Chapter 24.

Shamos, M. H., ed. *Great Experiments in Physics.* New York: Holt, Rinehart and Winston, 1959. Chapter 9.

Shapiro, Gilbert. *A Skeleton in the Darkroom.* San Francisco: Harper & Row, 1986. Chapter 2.

6. VACCINATION—EDWARD JENNER, A MILKMAID, AND SMALLPOX

"Jenner, Edward." *Encyclopaedia Britannica.* IX Edition. 1878. Vol. 13.

"Jenner, Edward." *Encyclopaedia Britannica.* 1962. Vol. 12.

Compere, E. L. "Research, Serendipity, and Orthopedic Surgery." *Journal of the American Medical Association* (21 December, 1957): P. 2070.

Clark, W. R. *The Experimental Foundations of Modern Immunology.* 3d ed. New York: John Wiley and Sons, 1986. Pp. 4–7.

7. DISCOVERIES OF CHEMICAL ELEMENTS

Accidental Scientific Discoveries. Chicago: Schaar and Co., 1955. This book offers articles on oxygen, iodine, and helium.

"Chemistry." *Encyclopaedia Britannica.* 1962. Vol. 5. A discussion of the phlogiston theory is given here in connection with the work of Priestley and Scheele. See also "Helium" (Vol. 11) and "Airship" (Vol. 1).

Oxygen

Farber, E., ed. *Great Chemists.* New York: Interscience, 1961. This book offers articles on Joseph Priestley, Carl Wilhelm Scheele, Joseph Louis Gay-Lussac, and Humphry Davy.

Moore, F. J. *A History of Chemistry.* 1st ed. New York: McGraw-Hill Co., 1918.

Priestley, Joseph. *Experiments and Observations on Different Kinds of Air.* Reprint No. 7. London: Alembic Club. This reprint gives the portions directly related to the discovery of oxygen.

Rhees, David J. "A Catalogue to an Exhibit Celebrating the 250th Birthday of Joseph Priestley and the Inauguration of the Center for History of Chemistry. Philadelphia: The Center for the History of Chemistry, 1983.

8. NITROUS OXIDE AND ETHER AS ANESTHETICS

Accidental Scientific Discoveries. Chicago: Schaar and Co., 1955. P. 17.

Farber, E., ed. *Great Chemists.* New York: Interscience, 1961.

Moore, F. J. *A History of Chemistry.* 1st ed. New York: McGraw-Hill Co., 1918.

Encyclopaedia Britannica. 1962. See "Anesthesia and Anesthetics" (Vol. 1) and "Davy, Sir Humphry" (Vol. 7).

9. WÖHLER'S SYNTHESIS OF UREA

"Wöhler, Friedrich." *Encyclopaedia Britannica.* 1962. Vol. 23.

Farber, E., ed. *Great Chemists.* New York: Interscience, 1961. P. 507.

Moore, F. J. *A History of Chemistry.* 1st ed. New York: McGraw-Hill Co., 1918. Chapter 12.

Williams, T. I., ed. *A Biographical Dictionary of Scientists.* 3d ed. New York: John Wiley and Sons, Inc., 1982. P. 566.

Wheland, G. W. *Advanced Organic Chemistry.* 2d ed. New York: John Wiley and Sons, Inc., 1949. P. 2.

Jaffe, Bernard. *Crucibles: The Story of Chemistry.* Premier Reprint. Greenwich, Conn.: Fawcett Publications, Inc., 1957.

10. DAGUERRE AND THE INVENTION OF PHOTOGRAPHY

Gernsheim, Helmut and Alison Gernsheim. *The History of Photography.* New York: McGraw-Hill Book Co., 1969. Chapter 6.

"Photography." *Encyclopaedia Britannica.* 1962. Vol. 17.

Schaaf, Larry J. "Nineteenth Century Photographic Processes." Report to the Faculty of the Graduate School of the University of Texas at Austin (August 1973). Chapters 2 and 3. Includes a definitive description of the daguerreotype.

11. RUBBER—NATURAL AND UNNATURAL

Vulcanization

Encyclopaedia Britannica. 1962. See "Rubber" (Vol. 19) and "Goodyear, Charles" (Vol. 10).

Halacy, D. S., Jr. *Science and Serendipity.* Philadelphia: Macrae Smith Co., 1967.

Garrett, A. B. *The Flash of Genius.* Princeton, N.J.: D. Van Nostrand Co., 1963.

Peirce, B. K. *Trials of an Inventor, Life and Discoveries of Charles Goodyear.* New York: Carlton and Porter, 1866. P. 106.

Synthetic Rubber

Berenbaum, M. B. *Encyclopedia of Polymer Science and Technology.* Vol. 11. New York: Wiley-Interscience, 1969. Pp. 425–447.

Gaylord, N. G., ed. *Polyethers.* New York: Wiley-Interscience, 1962. Chapter 13.

Marvel, C. S. "The Development of Polymer Chemistry in America—The Early Days." *Journal of Chemical Education* 58 (July 1981): 535.

12. PASTEUR: "LEFT-HANDED" AND "RIGHT-HANDED" MOLECULES

Kauffman, G. B., and R. D. Myers. "The Resolution of Racemic Acid." *Journal of Chemical Education* 53 (December 1975): 777.

"Pasteur, Louis." *Encyclopaedia Britannica.* 1962. Vol. 17.

Moore, F. J. *A History of Chemistry.* New York: McGraw-Hill Book Co., 1918. P. 204ff.

Vallery-Radot, Rene. *The Life of Pasteur.* Translated from French by Mrs. R. L. Devonshire. 2 vols. New York: McClure, Phillips and Co., 1902.

13. SYNTHETIC DYES AND PIGMENTS

Harrow, B. *Eminent Chemists of Our Time.* 2d ed. New York: D. Van Nostrand Co., 1927. Pp. 1, 241.

Accidental Scientific Discoveries. Chicago: Schaar and Co., 1955. P. 7.

Perkin, W. H. *Hofmann Memorial Lecture, Memorial Lectures delivered before the Chemical Society, 1893–1900.* London: Gurney and Jackson, 1901.

Crowther, J. G. *British Scientists of the Nineteenth Century.* London: Pelican Books; Penguin Books, 1940–1941. P. 375.

Levinstein, H. *Chemistry and Industry* 16 (1938): 1137.

Cronshaw, C. J. T. *Endeavour* 1 (1936): 79.

14. KEKULE: MOLECULAR ARCHITECTURE FROM DREAMS

Accidental Scientific Discoveries. Chicago: Schaar and Co., 1955. P. 10.

"Kekule." *Encyclopaedia Britannica.* 1962. Vol. 13.

Benfey, D. T. *Journal of Chemical Education* 35 (January 1958): 21.

Chemical and Engineering News (November 4, 1985): 22; (January 20, 1986): 3.

Weiss, U. and R. A. Brown. *Journal of Chemical Education* 64 (September 1987): 770.

Cannon, W. B. *The Way of an Investigator.* New York: W. W. Norton & Co., 1945. Chapter 5.

15. NOBEL: THE MAN, THE DISCOVERIES, AND THE PRIZES

"Nobel, Alfred Bernhard." *Encyclopaedia Britannica.* 1962. Vol. 16.

Bergengren, Erik. *Alfred Nobel, The Man and His Work.* 1st English ed. London: Thos. Nelson and Sons, Ltd., 1962.

Henriksson, F. *The Nobel Prizes and Their Founder, Alfred Nobel.* Stockholm: Alb. Bonniers, 1938.

Nobel Foundation, ed. *Nobel, The Man and His Prizes.* Amsterdam: Elsevier Publishing Co., 1962.

Halasz, N. *Nobel: A Biography of Alfred Nobel.* New York: Orion Press, 1959.

Sohlman, R. and H. Schuck. *Nobel, Dynamite and Peace.* New York: Cosmopolitan Book Corp., 1929.

Pauli, H. E. *Alfred Nobel, Dynamite King—Architect of Peace.* New York: L. B. Fischer, 1942.

16. CELLULOID AND RAYON: ARTIFICIAL IVORY AND SILK

"Celluloid." *Encyclopeadia Britannica.* 1962. Vol. 5.

Accidental Scientific Discoveries. Chicago: Schaar and Co., 1955. P. 43.

17. FRIEDEL AND CRAFTS—A LABORATORY ACCIDENT SPAWNS NEW INDUSTRIAL CHEMISTRY

Olah, G. A., and R. A. Dear. *Friedel-Crafts and Related Reactions,* edited by G. A. Olah. Vol. 1. New York: Interscience Publishers, 1963.

Roberts, R. M., and A. A. Khalaf. *Friedel-Crafts Alkylation Chemistry.* New York: Marcel Dekker, Inc., 1984.

18. HOW TO SUCCEED IN ARCHAEOLOGY WITHOUT REALLY TRYING

Fagan, Brian M. *The Adventure of Archaeology.* Washington, D.C.: National Geographic Society, 1985.

Digs That Produced Unexpected Results

Hao, Q., C. Heyi, and R. Suichu. *Out of China's Earth.* New York: Harry N. Abrams, Inc., and Beijing: China Pictorial, 1981. P. 65.

"Neanderthal Man." *Encyclopaedia Britannica.* 1962. Vol. 16.

Weaver, Kenneth F. "The Search for Our Ancestors." *National Geographic* (November 1985).

Austin American-Statesman (8, 9, 11 January 1985; 14 August 1987; 31 January 1988).

Washington Post (January 13, 1985).

Things That Just Turned Up

"Rosetta Stone." *Encyclopaedia Britannica.* 1962. Vol. 19.

Andrews, Carol. *The British Museum Book of the Rosetta Stone.* New York: Peter Bedrick Books, 1985.

Cottrell, Leonard, ed. *The Concise Encyclopaedia of Archaeology.* 2d ed. New York: Hawthorn Books, Inc., 1971. Pp. 323, 324.

Nature Sometimes Lends a Hand

Weaver, Kenneth F. "The Search for Our Ancestors." *National Geographic* (November 1985).

Boys and Caves

Tushingham, A. Douglas. "The Men Who Hid the Dead Sea Scrolls." *National Geographic* (December 1958).

Sponge Divers

Bass, George F. "Oldest Known Shipwreck Reveals Splendors of the Bronze Age." *National Geographic* (December 1987).

Katzev, Michael L. "Resurrecting the Oldest Known Greek Ship." *National Geographic* (June 1970).

Throckmorton, Peter. "Oldest Known Shipwreck Yields Bronze Age Cargo." *National Geographic* (May 1962).

______. "Thirty-three Centuries Under the Sea." *National Geographic* (May 1960).

Weaver, Kenneth F. "The Search for Our Ancestors." *National Geographic* (November 1985).

19. SOME ASTRONOMICAL SERENDIPITIES

Hannan, Patrick J., Rustum Roy, and John F. Christman. "Serendipity in Chemistry, Astronomy, Defense, and Other Useless Fields." *Chemtech* (July 1988): 402. Several other serendipitous discoveries, including others in astronomy, are described.

Shipman, Harry L. *Black Holes, Quasars, and the Universe.* 2d ed., Boston: Houghton Mifflin Co., 1980. P. 51ff.

Trefil, James S. *The Moment of Creation.* New York: Collier Books, Macmillan Publishing Co., 1983. Chapter 1.

20. ACCIDENTAL MEDICAL DISCOVERIES

Insulin

Cannon, Walter B. *The Way of an Investigator.* New York: W. W. Norton & Co., Inc., 1945. P. 72.

Sourkes, T. L. *Nobel Prize Winners in Medicine and Physiology.* London: Abelard-Schuman, 19. P. 111.

Encyclopaedia Britannica. 1962. See "Diabetes Mellitus" (Vol. 7); "Insulin" (Vol. 12); "Mering, Baron Joseph von" (Vol. 15), and "Minkowski, Oskar" (Vol. 15).

Allergy, Anaphylaxis, and Antihistamines

Sourkes, T. L. *Nobel Prize Winners in Medicine and Physiology.* London: Abelard-Schuman, 19. P. 79.

Cannon, Walter B. *The Way of an Investigator.* New York: W. W. Norton & Co., Inc., 1945. Pp. 71, 179.

Encyclopaedia Britannica. 1962. See "Allergy and Anaphylaxis" (Vol. 1); "Immunity" (Vol. 12); and "Richet, Charles Robert" (Vol. 19).

Nitrogen Mustards and Cancer Chemotherapy

Rhoads, C. P. *Journal of Mt. Sinai Hospital, N.Y.* 13 (1947): 299.

The Pill

Djerassi, Carl. "The Making of the Pill." *Science 84*. P. 127.

Russell E. Marker, personal communication, April 1988. Center for the History of Chemistry News Letter, Philadelphia (June 1987): 3.

LSD

Cohen, Sidney. *The Beyond Within: The LSD Story.* New York: Atheneum, 1970. Chapters 1 and 9.

Hofmann, Albert. "The Discovery and Subsequent Investigation on Naturally Occurring Hallucinogens." In "*Discoveries in Biological Psychiatry,*" edited by F. J. Ayd, Jr. and B. Blackwell. Philadelphia: Lippincott, 1970.

The Pap Test

Classics in Oncology 23 (May/June 1973).

Cameron, Charles S. "Dedication of the Papanicolaou Cancer Research Institute." *Journal of the American Medical Association* 182 (3 November 1962): 556.

Stenkvist, B., R. Bergstrom, G. Eklund, and C. H. Fox. "Papanicolaou Smear Screening and Cervical Cancer, What Can You Expect? *Journal of the American Medical Association* 252 (21 September 1984): 1423.

"Death of 'Dr. Pap'." *Medical World News* 2 (1962): 46.17.

Light and Infant Jaundice

Fincher, Jack. "Notice: Sunlight may be necessary for your health." *Smithsonian* (June 1985): 71.

Cholesterol Receptors

Banta, Bob. *Austin American-Statesman* (27 October 1985).

Brown, Michael S., and Joseph L. Goldstein. *Science* 232 (4 April 1986): 34.

Curtis, Gregory. *Texas Monthly* (December 1985).

Kolata, Gina. *Science* 221 (16 September 1983): 1164.

21. X RAYS, RADIOACTIVITY, AND NUCLEAR FISSION

Discovery of X rays by Röntgen

Accidental Scientific Discoveries. Chicago: Schaar and Co., 1955. P. 57.

Encyclopaedia Britannica. 1962. See "Rontgen, Wilhelm Conrad" (Vol. 19) and "X Rays" (Vol. 23).

Garrett, A. B. *The Flash of Genius.* Princeton, N.J.: D. Van Nostrand Co., Inc., 1963. P. 57.

Moore, F. J. A *History of Chemistry.* New York: McGraw-Hill, 1918. P. 252.

Nobel Foundation, ed. *Nobel, The Man and His Prizes.* Amsterdam: Elsevier Publishing Co., 1962. Pp. 446–448.

Shapiro, Gilbert. A *Skeleton in the Darkroom.* San Francisco: Harper and Row, 1986.

Discovery of Radioactivity by Becquerel

Accidental Scientific Discoveries. Chicago: Schaar and Co., 1955. P. 47.

"Becquerel." *Encyclopaedia Britannica.* 1962. Vol. 3.

Wolke, Robert L. "Marie Curie's Doctoral Thesis." *Journal of Chemical Education* 65 (1988): 561.

Artificial Radioactivity and Nuclear Fission

Lightman, Alan P. "To Cleave an Atom." *Science 84:* 103.

"Bohr, Niels." *Encyclopaedia Britannica.* 1962. Vol. 3.

22. SUBSTITUTE SUGAR: HOW SWEET IT IS—AND NON-FATTENING

Fieser, L. F. and M. *Advanced Organic Chemistry.* New York: Reinhold Publishing Co., 1961. P. 701.

Abstracts, *Journal of the Chemical Society* (1879): 628. A brief summary in English of the original article in the German journal (see the next reference).

Fahlberg, C., und I. Remsen. "Ueber die Oxydation des Orthotoluolsulfamids." *Berichte der Deutschen Chemischen Gesellschaft* 12 (1879): 469. This is the first publication that mentions the sweet taste of saccharin.

Journal of the Chemical Society (1927): 3186. Obituary notice for Ira Remsen.

O'Brien, L., and R. C. Gelardi. "Artificial Sweeteners." *Chemtech* (May 1981): 274.

Bakal, A. I. "Functionality of Combined Sweeteners in Several Food Applications." *Chemistry and Industry* (19 September 1983): 700.

Kirk-Othmer, ed. *Concise Encyclopedia of Chemical Technology.* New York: John Wiley and Sons, Inc., 1985. P. 1147.

23. SAFETY GLASS

"Glass." *Encyclopaedia Britannica.* IX Edition. 1878. Vol. 10.

"Glass." *Encyclopaedia Britannica.* 1962. Vol. 10.

Boyd, David C., and David A. Thompson. "Glass." *Encyclopedia of Chemical Technology.* Vol. 11. 3d ed. New York: John Wiley and Sons, 1982.

Lavin, Edward, and James A. Snelgrove. "Vinyl Polymers." *Encyclopedia of Chemical Technology.* Vol. 23. 3d ed. New York: John Wiley and Sons, 1982.

Sowers, Robert M. "Laminated Materials, Glass." *Encyclopedia of Chemical Technology.* Vol. 13. 3d ed. New York: John Wiley and Sons, 1982.

Wilson, J. "Safety Glass: Its History, Manufacture, Testing, and Development. *Journal of the Society of Glass Technology* 16 (1932): 67.

"Face-Saving Windshields." *DuPont Magazine* (September–October 1986): 20.

24. ANTIBIOTICS: PENICILLIN, SULFA DRUGS, AND MAGAININS

Penicillin: Fleming, Florey, and Chain

Ludovici, J. *Nobel Prize Winners.* London: Arco Publishers Ltd., 1957. P. 165.

Kauffman, G. B. "Maize, Melon and Mould—Keys to Penicillin Production." *Education in Chemistry* 17 (1980): 180.

Abraham, E. P. "Introduction." In *Chemistry and Biology of β-Lactam Antibiotics,* edited by R. B. Morin and M. Gorman. Vol. 1. New York: Academic Press, 1982.

Sourkes, T. L. *Nobel Prize Winners in Medicine and Physiology.* London: Abelard-Schuman, 1966. P. 235.

Garrett, A. B. *The Flash of Genius.* Princeton, NJ: D. Van Nostrand Co., Inc., 1963. P. 31.

Halacy, D. S. Jr. *Science and Serendipity.* Philadelphia: Macrae Smith Co., 1967. P. 31.

Fieser, L. F. and M. *Topics in Organic Chemistry.* New York: Reinhold Publishing Corp., 1963. Pp. 309, 313ff.

Accidental Scientific Discoveries. Chicago: Schaar and Co., 1955. P. 35.

Sulfa Drugs: Domagk, Fourneau, and the Trefouels

Northey, E. H. *The Sulfonamides and Allied Compounds.* American Chemical Society Monographs. New York: Reinhold Publishing Corp., 1948. P. 1ff.

Sourkes, T. L. *Nobel Prize Winners in Medicine and Physiology.* London: Abelard-Schuman, 1966. P. 214.

Silverman, M. *Magic in a Bottle.* New York: Macmillan: 1941. P. 283ff.

Noller, C. R. *Chemistry of Organic Compounds.* 3d ed. Philadelphia: W. B. Saunders Co., 1965. P. 534.

Solomons, T. W. G. *Fundamentals of Organic Chemistry.* 2d ed. New York: John Wiley and Sons, 1986. P. 743.

"Domagk, Gerhard." *Encyclopaedia Britannica.* 1962. Vol. 7.

Magainins: Antibiotics from a Frog's Skin

Okie, Susan (Washington Post Service). *Austin American-Statesman* 30 July 1987).

Chemical and Engineering News (3 August 1987): 22.

25. NYLON: COLD DRAWING DOES THE TRICK

Marvel, C. S. "The Development of Polymer Chemistry in America—The Early Days." *Journal of Chemical Education* 58 (July 1981): 535.

Farber, E., ed. *Great Chemists.* New York: Interscience, 1961. P. 1600.

26. POLYETHYLENE: THANKS TO LEAKY AND DIRTY EQUIPMENT

Swallow, J. C. "The History of Polythene." *Polythene—The Technology and Uses of Ethylene Polymers,* edited by A. Renfrew. 2d ed. London: Iliffe and Sons, Ltd., 1960.

Chien, J. C. W. *Coordination Polymerization, A Memorial to Karl Ziegler.* New York: Academic Press, Inc., 1975.

Friedrich, M. E. P., and C. S. Marvel. "The Reaction Between Alkali Metal Alkyls and Quaternary Arsonium Compounds." *Journal of the American Chemical Society* 52 (1930): 376. This article briefly mentions the polymerization of ethylene by butyl lithium.

Owen, E. D., ed. *Degradation and Stabilization of PVC.* London: Elsevier Applied Science Publishers, 1984. Chapter 1.

Myerly, Richard C. *Journal of Chemical Education* 57 (1980): 437

Price, Charles C. et al. *Journal of the American Chemical Society* 78 (1956): 690.

27. TEFLON: OUT OF THE ATOM BOMB AND INTO THE FRYING PAN

The American Institute of Chemists. "The Original Teflon Man." *The Chemist* (May 1985): 4.

Midgely, T., Jr. The Perkin Medal address. *Journal of Industrial and Engineering Chemistry* 29 (1937): 241.

Plunkett, Roy J. "The History of Polytetrafluoroethylene Discovery and Development." Paper given at the AAAS Meeting in Philadelphia, May 1986.

28. GASOLINE TECHNOLOGY: FLOWERY THEORIES AND GAS TO GASOLINE

Robert, Joseph C. *Ethyl, A History of the Corporation and the People Who Made It.* Charlottesville, Va.: University of Virginia Press, 1983. P. 104ff.

Chemical and Engineering News (22 June 1987).

29. DRUGS ACCIDENTALLY FOUND GOOD FOR SOMETHING ELSE

Aspirin

Badger, G. M. In *Proceedings of the Royal Australian Chemical Institute.* October 1973. P. 273.

Van, Jon (Chicago Tribune Service). *Austin American-Statesman* (February 16, 1988): C9.

Psychoactive Drugs

Snyder, Solomon H. *Medicated Minds.*" *Science 84.* P. 141.

Center for the History of Chemistry News Letter, Philadelphia (Fall 1986): 6.

Antiarrhythmic Drugs So Things Won't Go Bump in the Night

Katz, R. L., and G. J. Katz. In *Lidocaine in the Treatment of Ventricular Arrhythmias: Proceedings of a Symposium held in Edinburgh in September 1970,* edited by D. B. Scott and D. G. Julian. Edinburgh: E. and S. Livingstone, 1971. P. 112.

Burstein, Charles. "Treatment of Acute Arrhythmias During Anesthesia by Intravenous Procaine." *Anesthesiology* 7 (1 March 1946): 113.

Minoxidil: A Hair-Raising Experience

Steinbrook, Robert (Los Angeles Times Service). *Austin American-Statesman* (21 September 1986).

Rumsfield, J. A., D. P. West, and V. Fiedler-Weiss. "Topical Minoxidil Therapy for Hair Regrowth." *Clinical Pharmacy* 6 (May 1987): 386.

Stern, Robert S. "Topical Minoxidil." *Archives of Dermatology* 123 (January 1987): 62.

"Topical Minoxidil for Baldness," *The Medical Letter on Drugs and Therapeutics* 29 (September 1987): 87.

Interferon: Cancer and Arthritis

Taylor-Papadimitriou, Joyce, ed. *Interferons: Their Impact in Biology and Medicine.* Oxford: Oxford University Press, 1985.

"Interferon found to relieve arthritis." *Austin American-Statesman* (AP) (October 9, 1985).

30. DRUGS FROM SEWAGE AND DIRT

Cephalosporins

Abraham, E. P. "Introduction." In *Chemistry and Biology of β-Lactam Antibiotics,* edited by R. B. Morin and M. Gorman. Vol. 1. New York: Academic Press, 1982.

Manhas, M. S., and A. K. Bose. *β-Lactams: Natural and Synthetic.* New York: Wiley-Interscience, 1971. Part 1, pp. 31–49.

Research papers by E. P. Abraham and co-workers in the *Biochemical Journal,* Vol. 50, p. 168 (1951); Vol. 58, p. 94 (1954); Vol. 79, p. 377 (1961).

Gregory, G. I., ed. *Recent Advances in the Chemistry of β-Lactam Antibiotics: Proceedings of the Second International Symposium, Cambridge, England.* London: The Royal Society of Chemistry, 1981.

Hodgkin, D. C., and E. N. Maslen. *Biochemical Journal* 79 (1961): 393.

Brown, A. G., and S. M. Roberts eds. *Recent Advances in the Chemistry of β-Lactam Antibiotics: Proceedings of the Third International Symposia, Cambridge, England.* London: The Royal Society of Chemistry, 1985.

Glatt, A. E. "Third-Generation Cephalosporins." *Physician Assistant* (January 1986).

Cyclosporine

Borel, J. F. In *Cyclosporin A: Proceedings of an International Conference on Cyclosporin A, Cambridge, England,* edited by D. J. G. White. Amsterdam: Elsevier Biomedical Press, 1982.

Kolata, Gina. "Drug Transforms Transplant Medicine." *Science* (July 1983): 40.

31. BROWN AND WITTIG: BORON AND PHOSPHORUS IN ORGANIC SYNTHESIS

Hydroboration

Brown, H. C. *Hydroboration.* New York: W. A. Benjamin, Inc., 1962. Chapter 2.

Brown, H. C. *Boranes in Organic Chemistry.* Ithaca, N.Y.: Cornell University Press, 1972. P. 258.

Loudon, G. Marc. *Organic Chemistry.* Reading, Mass.: Addison-Wesley Publishing Co., 1984. Pp. 916, 917.

The Robert A. Welch Foundation. *Research Bulletin N. 33 and Program of the XVII Conference on Chemical Research* (5–7 November 1973). Houston, Texas.

Synthesis of Alkenes

March, J. *Advanced Organic Chemistry.* 2d ed. New York: Wiley-Interscience, 1985. P. 845.

Wittig, G., and G. Geissler. "The Course of Reactions of Pentaphenylphosphorus and Certain Derivatives." *Annalen der Chemie* 580 (1953): 44.

Wittig, G., and U. Schoellkopf. "Triphenylphosphinemethylene as an Olefinforming Reagent." *Chemische Berichte* 97 (1954): 1318.

Wittig, G. "Variationen zu einem Thema von Staudinger; ein Beitrag zur Geschichte der Phosphororganischen Carbonylolefinierung." *Pure and Applied Chemistry* 9 (1964): 245. Contains a summary in English.

"Nobel Prizes." *Encyclopaedia Britannica, Yearbook for 1980.* P. 101.

32. POLYCARBONATES: TOUGH STUFF

Fox, D. W. "From Plant Pigments to Polymers." *The Chemist* (September 1987): 12. This is the text of Dr. Fox's 1987 Chemical Pioneer Address.

Fox, D. W. "Polycarbonates." In *Encyclopedia of Chemical Technology.* Vol. 18. 3d ed. New York: John Wiley and Sons, 1982. P. 479.

33. VELCRO AND OTHER GIFTS OF SERENDIPITY TO MODERN LIVING

Velcro: From Cockleburs to Spaceships

Brochures published by Velcro USA, Inc.

"Cocklebur." *Encyclopaedia Britannica.* 1962. Vol. 5.

Ivory Soap; Corn and Wheat Flakes

Fichter, George. *United Airlines Magazine* 31 (October 1986): 87, 90.

Post-its

Hutchinson, Julie. *Austin American-Statesman* (1 September 1985).
Brochure available from the 3M Company, courtesy of Rich Sanders in the Technical Services Department and Judy Boroski, Staff Marketing Services, St. Paul, Minn.

34. DNA: THE COIL OF LIFE

Watson, James D. *The Double Helix,* New York: Atheneum, 1968.

Watson, James D., and F. H. C. Crick. "A Structure for Deoxyribose Nucleic Acid." *Nature* 171 (1953): 737, 748.

35. CONCEPTIONS, MISCONCEPTIONS, AND ACCIDENTS IN ORGANIC SYNTHESIS

Encyclopaedia Britannica, Yearbook for 1970. See "Barton, Derek H. R." (p. 138) and "Hassel, Odd," (p. 149).

Barton, D. H. R. In *Stereochemistry of Organic and Bioorganic Transformations,* edited by W. Bartmann and K. B. Sharpless. Weinheim, Federal Republic of Germany, VCH Verlagsgesellschaft mbH, 1987. P. 205.

Barton, D. H. R. *Experientia* 6 (1950): 316.

Barton, D. H. R. "The Invention of Organic Reactions Useful in Bioorganic Chemistry." In *Frontiers in Bioorganic Chemistry and Molecular Biology,* edited by Yu. A. Ovchinnikov and M. N. Kolosov. Elsevier/North-Holland Biomedical Press, 1979.

Barbier, M., D. H. R. Barton, M. Devys, and R. S. Topgi. "A Simple Synthesis of the Tropone Nucleus." *Journal of the Chemical Society, Chemical Communications* (1984): 743.

36. CHEMICAL CROWNS AND CRYPTS

"Winners of Nobels in Chemistry and Physics." *New York Times* (October 19, 1987).

"Nobel Prizes: U. S., French Chemists Share Award." *Chemical and Engineering News* (19 October 1987): 4.

"Progress Made in Synthesizing Enzyme Mimics." *Chemical and Engineering News* (19 October 1987): 30.

"Chemistry in the Image of Biology." *Science* (30 October 1987): 611.

Du Pont Magazine, College Report supplement (January–February 1988).

Industrial Chemist (February 1988): 22. An interview with Charles J. Pedersen.

Pedersen, C. J. "Cyclic Polyethers and Their Complexes with Metal Salts." *Journal of the American Chemical Society* 89 (1967): 7017.

Pedersen, C. J. "The Discovery of the Crown Ethers" (Nobel lecture). *Angewandte Chemie,* International Edition *in English* 27 (1988): 1021.

EPILOGUE

Lenox, Ronald S., *Journal of Chemical Education* 64(1985):282.

NAME INDEX

Names in **Bold Face** type are those of Nobel Prize laureates.

SUBJECT INDEX